The Institute of Biology's
Studies in Biology no. 48

Plants and Mineral Salts

James F. Sutcliffe
D.Sc., Ph.D., F.I.Biol.
Professor of Plant Physiology, University of Sussex

Dennis A. Baker
B.Sc., Ph.D., M.I.Biol.
Reader in Biology, University of Sussex

Edward Arnold

First published 1974
by Edward Arnold (Publishers) Ltd.,
25 Hill Street, London W1X 8LL

Boards edition ISBN 0 7131 2451 2
Paper edition ISBN 0 7131 2452 0

Printed in Great Britain by
The Camelot Press Ltd, Southampton

General Preface to the Series

It is no longer possible for one textbook to cover the whole field of Biology and to remain sufficiently up-to-date. At the same time teachers and students at school, college or university need to keep abreast of recent trends and know where significant developments are taking place.

To meet the need for this progressive approach the Institute of Biology has for some years sponsored this series of booklets dealing with subjects specially selected by a panel of editors. The enthusiastic acceptance of the series by teachers and students at school, college and university shows the usefulness of the books in providing a clear and up-to-date coverage of topics, particularly in areas of research and changing views.

Among features of the series are the attention given to methods, the inclusion of a selected list of books for further reading and, wherever possible, suggestions for practical work.

Readers' comments will be welcomed by the authors or the Education Officer of the Institute.

1974

The Institute of Biology,
41 Queens Gate,
London, SW7 5HU

Preface

In addition to carbon dioxide and water, plants require a variety of mineral elements which they obtain from the soil. The processes by which ions are absorbed, transported, utilized and sometimes excreted by plants have been the subject of extensive and intensive research over the last fifty years and an enormous literature has been built up. Ion transport, although of such great biological importance, is an aspect of plant physiology which often receives cursory treatment in elementary textbooks, while the comprehensive monographs and reviews written for the specialist are not always readily available and are often difficult and time-consuming to read.

The aim of the present book is to present in a clear and concise form an outline of the mineral nutrition of plants in all its facets, including plant-soil relations, functions of the essential elements, mechanisms of uptake by cells, cell to cell transport, distribution in the xylem and phloem and the effects of high salinity on plant growth. In particular, we have attempted to integrate the more traditional aspects of mineral nutrition with the current biophysical approach to ion uptake.

Our thanks are due to Professor J. F. Loneragan of the University of Western Australia, for helpful advice concerning the lay-out and content of Chapter 2; to Dr E. J. Hewitt for supplying photographs of mineral deficiency symptoms; and to our colleagues in the Plant Physiology group at the University of Sussex, for stimulating discussions. We are grateful to Miss Nora Browning who typed the final draft of the manuscript.

Sussex, 1974

J.F.S. and D.A.B.

Contents

1 Salt Supply

1.1 Sources of mineral salts

Plants feed so unobtrusively that it took many centuries of observation and investigation before the individual roles of light, air and soil were properly understood. The processes whereby green plants in the light convert carbon dioxide from the air and inorganic salts from the soil into organic substances are just as important to animals as to plants because even carnivores depend on photosynthesis as an ultimate source of food.

Most plants obtain the bulk of the mineral elements they require via the roots, but sometimes salts are taken up partly through the leaves. In submerged aquatic plants they are absorbed over the whole plant surface and the roots are of minor significance in this connection. Epiphytes, growing on the branches and trunks of trees, and the carpets of bryophytes on a woodland floor also depend mainly on the leaves for absorption of inorganic nutrients leached by rain from the foliage above. Parasites and hemi-parasites develop special absorptive organs (haustoria) by means of which they extract mineral salts from the vascular tissues of their host.

Mineral salts originate from rocks of the lithosphere by the process of weathering whereby complex crystalline structures are broken down slowly by physical and chemical processes to soluble compounds. These dissociate to a greater or less extent in water forming positively-charged

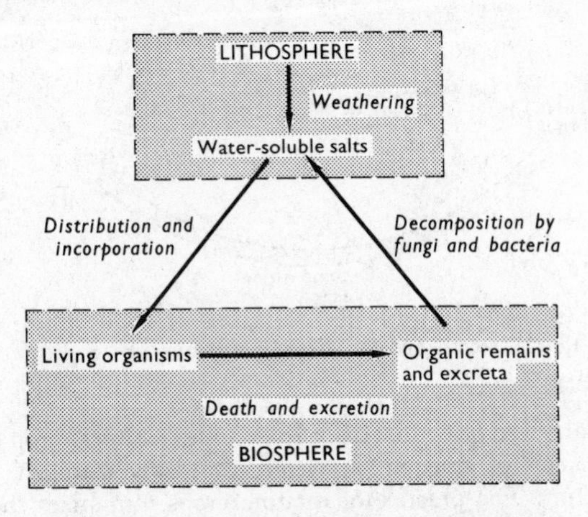

Fig. 1–1 Circulation of mineral elements

cations, e.g. potassium (K^+), calcium (Ca^{2+}) and iron (Fe^{2+} and Fe^{3+}), and negatively charged anions, such as chloride (Cl^-), sulphate ($SO_4{}^{2-}$) and phosphate ($H_2PO_4{}^-$, $HPO_4{}^{2-}$, $PO_4{}^{3-}$). They are carried away by rain, streams and rivers into soils, lakes and oceans where some are absorbed and incorporated in the biosphere. Inorganic substances used in this way are recycled by breakdown of waste products and organic remains of animals and plants through the action of fungi and bacteria (Fig. 1–1).

There is one important element, *nitrogen*, which does not occur in rocks. Its presence in the soil and in natural waters in the forms of nitrite ($NO_2{}^-$), nitrate ($NO_3{}^-$) and ammonium ($NH_4{}^+$) ions depends mainly on the decomposition of nitrogenous organic compounds—especially plant and animal proteins, and to a lesser extent on fixation of atmospheric nitrogen by electrical discharges and by the activity of micro-organisms (Fig. 1–2).

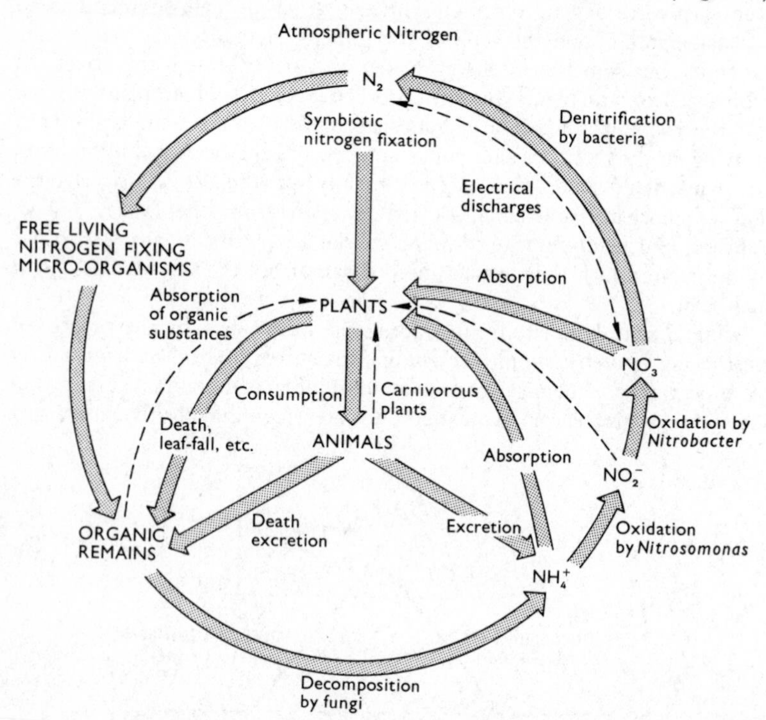

Fig. 1–2 The nitrogen cycle. Minor pathways are indicated by dashed arrows

When plants are grown intensively as an agricultural crop they rapidly deplete the soil of essential nutrients, especially nitrogen, phosphorus and potassium. The practice of rotating crops minimizes this effect because different species have somewhat different nutrient requirements, but even so manuring is necessary to maintain high yields. Artificial fertilizers, especially those containing nitrogen, phosphorus, potassium and calcium, are extensively used nowadays for this purpose.

1.2. Composition of plant ash

When plant material is burnt in air, the organic matter is destroyed and a residue of inorganic salts, the *ash*, remains. During burning, elements present in the original tissues whether as constituents of organic molecules, as crystalline deposits of insoluble salts, or in ionic form are converted mainly to their oxides. These can be dissolved in a strong acid, such as concentrated hydrochloric acid, and the resulting solution, after dilution, may be analysed by chemical or physical methods. Such analyses reveal a large number of constituents (Fig. 1–3) but little or no nitrogen

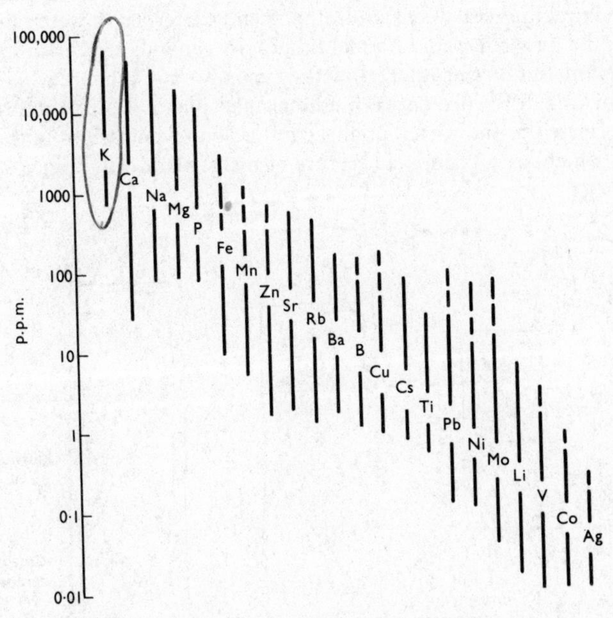

Fig. 1–3 Range of concentration (parts per million of dry matter) of various elements in plant ash. Dashed lines indicate exceptional values. (redrawn from MITCHELL, 1954)

is found because it is converted to ammonia and gaseous oxides which are lost during combustion. Other elements, notably phosphorus and sulphur, may be lost in a similar way, but to a lesser extent. To minimize such losses a procedure called 'wet ashing', which involves digestion of the dried material with perchloric and hydrochloric acids at controlled temperatures is commonly used in research.

The most convenient method of measuring the amounts of such elements as potassium, sodium, calcium and magnesium in solutions is by the use of emission or atomic absorption flame spectrophotometry. Sensitive spectroscopic analysis reveals the presence of a great many elements in plant extracts, some of which occur at extremely low concentration (Fig. 1–3). Radium, one of the rarer natural elements, has been detected in some woody angiosperms at a concentration of only 10^{-9} parts per million of dry matter.

The range of concentration of a particular element varies widely between different plants and is also affected by the conditions under which the plants are grown (see below), but all plant ash is similar in certain respects. The main constituent is usually *potassium*, which often comprises nearly 50% of the total weight of ash. The name 'potasssium', derived from 'pot ashes', commemorates an early method of preparing potassium salts by extracting sea-weeds and other plant materials in water and evaporating the liquor in pans. Animal tissues, in general, are much less rich in potassium, but on the other hand, they usually contain more *sodium*. The reason for this difference between animals and plants is still obscure. It is possibly related to the preferential accumulation of potassium in the large vacuoles which are a prominent feature of many plant cells (Fig. 1–4), but

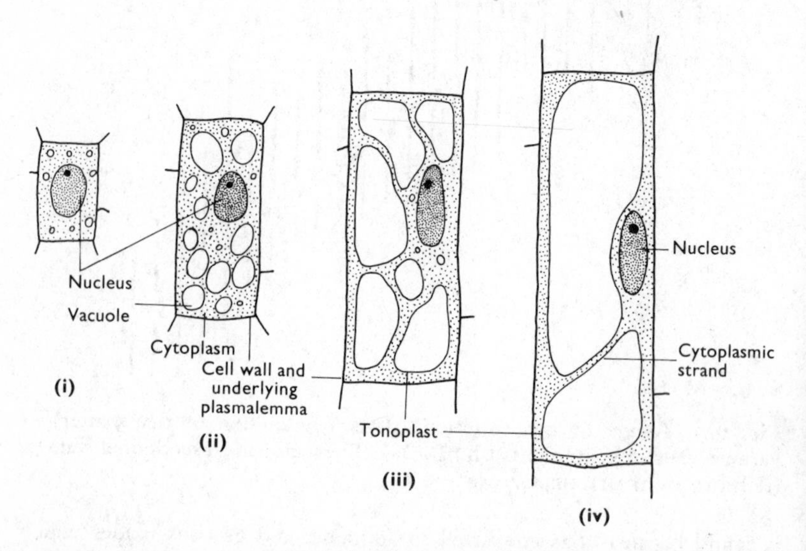

Fig. 1–4 Stages in development of a parenchyma cell in the elongating zone of a root (i), a meristematic cell; (ii–iii), stages in vacuolation; (iv), a fully-vacuolated cell. (Drawings by C. MORTLOCK)

meristematic cells, which only have small vacuoles, show an even greater preference for potassium over sodium (SUTCLIFFE and COUNTER, 1962). It is evident that during evolution animals have maintained an internal environment more similar to that of the sea than have most plants. One consequence of this is that the diet of domesticated herbivorous animals such as cattle, must be supplemented with common salt.

After potassium, *calcium* is often the most abundant element in plant ash, although its concentration ranges widely from trace amounts in maize (*Zea mays*) grains to over 7% of the dry weight in mature sunflower (*Helianthus annuus*) leaves. In contrast to potassium, calcium occurs mainly in combined forms either associated with cell walls or as crystalline deposits of insoluble calcium salts, such as calcium oxalate, in the cytoplasm. Plant ash is also rich in *magnesium*, which is a constituent of some organic molecules, including chlorophyll, and also occurs as free ions in the cell sap. *Phosphorus* is present in plant ash mainly as phosphorus pentoxide produced by oxidation of organic and inorganic phosphates during burning. The ash from plant seeds is particularly rich in phosphorus, most of which is derived from phytin (calcium and magnesium myo-inositol hexaphosphate), a common form of storage phosphate.

1.3 Genetic control of salt content

Individual plant species and varieties differ markedly in their salt content even when they are grown under the same conditions. Sometimes these differences can be traced to differences in the size and form of the root system which enables the plants to exploit different regions of the soil, but in other cases they appear to be attributable to differing characteristics of absorption *per se*. COLLANDER (1941) showed that when two plantains, *Plantago lanceolata* and *Plantago maritima*, were grown in the same nutrient solution containing equal concentrations of potassium and sodium ions, the long-leaved plantain absorbed more than twice as much potassium as sodium, whereas the sea plantain took up rather more sodium than potassium. Grasses tend to have lower amounts of calcium than leguminous plants both in the field and when grown in solution culture, but on the other hand they are characteristically rich in *silicon*. This element which comprises 1% or more of the dry weight of maize plants is mainly present as crystals of silica (opals) and as deposits in cell walls. In one group of algae, the diatoms, the cell wall is composed entirely of silicious material, and the ash of plankton, when rich in diatoms, may contain as much as 20% of silicon. Whereas most plants contain very little *aluminium*, large amounts occur in the Club-mosses (*Lycopodium spp.*) and in some species of one angiosperm family, the Diapensiaceae. This accounts for the use in earlier times of these plants as mordants in the dyeing industry. Some members of the leguminous genus *Astragalus* accumulate *selenium* in exceptional amounts and when growing on selenium-rich soils they contain so much that animals eating them may suffer from selenium poisoning ('alkali disease'). Such differences between different groups of plants presumably have a genetic basis. WEISS (1943) did breeding experiments with two varieties of soya bean (*Glycine max*), which take up iron at different rates, and he was able to show that in this plant the ability to absorb iron from low concentrations is controlled by a single gene. A few more instances of monogenic control of absorption of specific elements have been reported more recently, but in most cases it appears that several gene loci are involved and the inheritance of absorptive capacity is therefore usually quite complex.

1.4 Influence of the environment

Within the limits set by its genetic constitution the salt content of a plant is controlled by environmental factors of which *nutrient supply* is often of paramount importance. It is likely that a plant absorbs, at least in small amounts, every element presented to it; some which do not occur naturally, such as plutonium produced in nuclear reactors, can be taken up by plants. When more of a particular element is provided its concentration in the plant usually increases and the levels of some other elements fall. It was shown, for example, by research workers at Rothamsted Experimental Station near Harpenden, where some of the earliest scientific investigations of mineral nutrition of crops were made, that application of potassium as a fertilizer to grassland swards causes an increase in the amount of potassium in the ash and reduces the levels of other constituents (Table 1).

Table 1 Concentration of some elements in the ash from herbage of grass plots at Rothamsted supplied with nitrogen, phosphorus and potassium (N.P.K.) or with nitrogen and phosphorus only (N.P.) (Data from RUSSELL, 1974)

Element	*Parts per million of dry matter* *N.P.*	*N.P.K.*
K	8 112	11 372
Na	4 370	2 070
Ca	6 500	4 284
P	3 200	1 612

Similarly, high concentrations of calcium in the soil usually cause a reduction in the levels of potassium, sodium and phosphorus. High calcium content of soil is usually accompanied by alkalinity, which itself has a marked effect on the salt content of plants. Some elements, e.g. iron and manganese, are taken up less readily at high pH due to conversion of easily-absorbed ferrous (Fe^{2+}) and manganous (Mn^{2+}) ions to more highly oxidized, less available ferric (Fe^{3+}) and manganic (Mn^{3+}) ions. The failure of some 'calcifuge' plants, such as rhododendrons, to grow well in calcareous soils is attributable to an inability to absorb sufficient iron or manganese under these conditions. The problem can be overcome by supplying the elements in a chelated form, e.g. as iron ethylene diamine-tetra-acetic acid (Fe-EDTA). Phosphate absorption is sensitive to pH because the univalent ion ($H_2PO_4^-$) which predominates in acid solution is more readily absorbed than either the bivalent or trivalent form.

Analyses of vacuolar sap extracted from large cells of certain algae, e.g. *Nitella clavata* and *Valonia macrophysa*, show that the concentration of various cations and anions in the sap is many times higher than that of the pond water or sea water in which the plant grows (Fig. 1–5). This

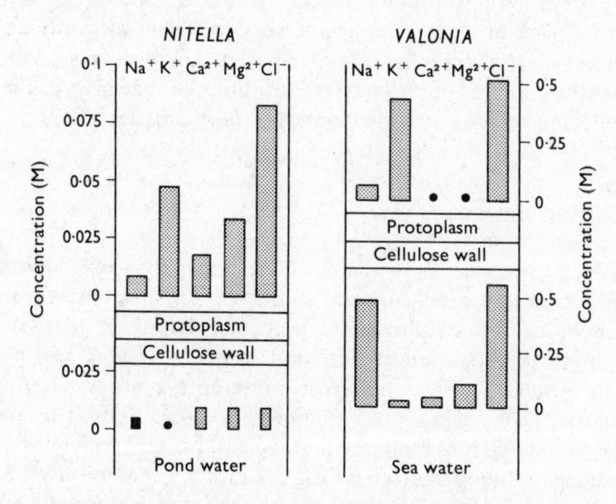

Fig. 1–5 Diagram showing the relative concentrations of various ions in the sap of two algae, *Nitella clavata* and *Valonia macrophysa* and in the medium in which the plants were grown. (Redrawn from HOAGLAND, 1944)

fact, together with the observation mentioned above that plants absorb ions selectively, led to the conclusion that absorption is an 'active process' dependent on metabolic energy, even before rigorous proof was obtained. Further evidence supporting this view comes from the demonstration that uptake is affected by such factors as *temperature* and *aeration* which also influence metabolism (see Chapters 3 and 4).

1.5 Salt absorption and growth

Since growth is affected by the same environmental factors as influence salt absorption, it is not surprising that there is a close correlation between salt content and growth. The phase of rapid vegetative growth, whether of a whole plant or of a cell, is accompanied by a large increase in salt content, and absorption declines when growth slows down. In the early stages of seedling growth salt content often rises more rapidly than dry matter, but later the situation is reversed so that even though total salt content continues to rise, the amount expressed in parts per million of dry matter begins to decline. In a similar way changes in the concentration of an ion in the vacuolar sap of a growing cell are dependent both on the rate of absorption of salt and on the concomitant absorption of water. During the rapid phase of extension growth in a root cell the concentration of ions in the cell sap often falls temporarily because absorption of water occurs even more rapidly than absorption of salt. Great care must be taken therefore, when interpreting changes in salt content expressed as a proportion of dry matter (p.p.m.) or per unit of water content (e.g. mM).

Different parts of a plant may have very different salt contents and the amounts present in different organs alter markedly during their growth. This is attributable mainly to the redistribution of materials in the phloem from senescing to actively growing organs (see Chapter 5).

1.6 Solution culture

It is well known that most plants can be grown perfectly well with their roots immersed in aerated nutrient solutions. This technique now known as 'hydroponics' was employed by John Woodward in the seventeenth century in an investigation of nutrition in mint (*Mentha* sp.) plants and was used extensively by the great nineteenth-century German plant physiologists, Sachs, Knop and Pfeffer (Fig. 1–6). Besides its usefulness in studies of mineral nutrition, the method also has commercial applications and nowadays a number of crops can be grown economically in solution culture. To support the plant mechanically the roots are usually embedded in an inert material, such as vermiculite (hydrated magnesium-aluminium silicate), perlite, plastic or glass beads on to which the nutrient solution is poured. The solution is either replaced at intervals as it becomes depleted or is allowed to flow continuously over the roots. For technical details of methods of solution culture, see HEWITT (1966).

For optimal growth the nutrient solution must contain all the elements which the plant requires, in a suitable form and in approximately the right proportions. The composition of several solutions in which many plants grow well is given in Table 2. No one solution is best for all plants and HOMÈS (1963) has devised a systematic procedure for varying the concentration of major nutrients to obtain the right proportions for a particular plant. The pH value of the solution is also critical and is usually adjusted to an optimal value (often about pH 6.5) by addition of acid (HCl) or alkali (NaOH) as required.

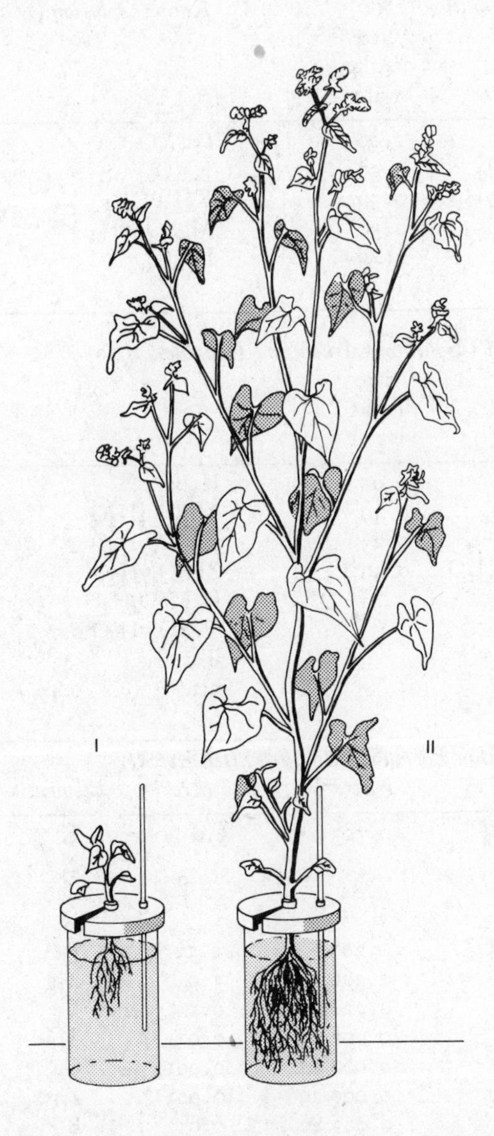

Fig. 1–6 Buckwheat grown in solution culture: I, nutrient solution without potassium; II, complete nutrient medium. (After PFEFFER, W., 1900, *The Physiology of Plants*, edited and translated by A. J. Ewart, Clarendon Press, Oxford.)

Table 2 The composition of some nutrient solutions suitable for higher plants

A. *Sachs' Solution (1860)*

	g $litre^{-1}$ distilled water
KNO_3	1.00
$Ca_3(PO_4)_2$	0.50
$MgSO_4.7H_2O$	0.50
$CaSO_4$	0.50
NaCl	0.25
$FeSO_4$	Trace

B. *Knop's Solution (1865)*

	g $litre^{-1}$ distilled water
$Ca(NO_3)_2$	0.8
KNO_3	0.2
KH_2PO_4	0.2
$MgSO_4.7H_2O$	0.2
$FePO_4$	0.1

C. *Hoagland's Solution (Arnon and Hoagland* 1940)

	g $litre^{-1}$ distilled water		mg $litre^{-1}$ distilled water
KNO_3	1.02	H_3BO_3	2.86
$Ca(NO_3)_2$	0.49	$MnCl_24H_2O$	1.81
$NH_4H_2PO_4$	0.23	$CuSO_45H_2O$	0.08
$MgSO_47H_2O$	0.49	$ZnSO_47H_2O$	0.22
		$H_2MoO_4H_2O$	0.09
		$FeSO_47H_2O$ 0.5%	0.6ml per litre added 3 × weekly
		Tartaric acid 0.4%	

D. *Long Ashton Formula (Modified after Hewitt)*

Salt	*g* $litre^{-1}$	*mM* $litre^{-1}$	*Element*	*p.p.m.*
KNO_3	0.505	5.0	K	195
			N	70
$Ca(NO_3)_2$	0.82	5.0	Ca	200
			N	140
$NaH_2PO_42H_2O$	0.208	1.33	P	41
$MgSO_47H_2O$	0.369	3.0	Mg	24
Ferric citrate	0.0245	0.1	Fe	5.6
$MnSO_4$	0.002 23	0.01	Mn	0.55
$CuSO_45H_2O$	0.000 24	0.001	Cu	0.064
$ZnSO_47H_2O$	0.000 296	0.001	Zn	0.065
H_3BO_3	0.001 86	0.033	B	0.37
$(NH_4)_6Mo_7O_{24}4H_2O$	0.000 035	0.000 2	Mo	0.019
$CoSO_47H_2O$	0.000 028	0.000 1	Co	0.000 6
NaCl	0.005 85	0.1	Cl	3.55

2 Salt Requirement

2.1 Essential and non-essential elements

Presence of a particular element in plant ash does not necessarily mean that it has any useful function. In order to prove that an element is essential it is necessary to show that a plant does not grow normally and complete its life cycle unless a certain minimum amount of the element is supplied. Occasionally, the omission of an element from the culture medium interferes with growth indirectly, perhaps by preventing the absorption of another element required by the plant or by promoting the uptake of a toxic substance. According to ARNON (1950) such a *beneficial* element should not be considered essential unless it also has a specific function in the plant. Because it has one or more unique functions an essential element is one which cannot usually be replaced by any other element, but there are a few cases known in which a particular function may be carried out by two related elements. Potassium and rubidium are apparently interchangeable in many bacteria and in the sea-lettuce, *Ulva lactuca*, while strontium can replace calcium in the alga, *Chlorella*, and possibly in some fungi. In such cases the commoner of the two elements is the one which is considered to be essential because it is the one most likely to be functional in nature.

Another difficulty arises if an element is shown to be required only under certain conditions. Molybdenum, for example, appears to be essential for the alga *Scenedesmus* only if nitrogen is supplied solely as nitrate. It is clearly needed for the metabolism of nitrate and therefore if nitrogen is available in other forms, e.g. as ammonia or as urea, molybdenum is not required. It has been established that the molybdenum requirement of legumes is much higher when they are fixing atmospheric nitrogen than when they are utilizing nitrate or ammonia. This appears to be related to involvement of molybdenum in the nitrogen-fixing enzyme, nitrogenase. Legumes that fix nitrogen also need the element cobalt, which is not otherwise known for certain to be an essential element in plants although it is added sometimes to culture solutions (Table 2 D).

It is possible to examine in detail the nutrient requirements of a given plant species under various conditions, but because of the immense amount of labour involved it will never be feasible to establish that a particular element is universally essential to all plants under all possible conditions. It is estimated that there are about 200 000 species of flowering plants alone, and the mineral requirements of less than 100 have been studied in detail. If an element has been found to be necessary for normal growth of a diverse range of plants and is known to participate in what is believed to be a universal physiological process, it is reasonable to conclude that it is required by all, even though this may never be proved.

2.2 The essential elements

The study of the mineral salt requirement of plants involves the careful control and manipulation of nutrients in the rooting medium. To make a soil sufficiently free of any particular element is impossible and therefore most investigators use sand or solution culture. Solution culture provides an excellent means of controlling the form, quantity and relative proportions of various elements. Any element can be almost completely excluded from the solution if precautions are taken to avoid contamination by using glass distilled water, highly purified chemicals and clean containers of pyrex glass or plastic. Even dust in the surrounding air is a possible source of contamination.

The procedure for testing whether an element is essential usually involves growing plants from seed in solutions which lack only that element and comparing them with plants grown in a complete nutrient medium. However, it is not possible to omit one element from a culture solution without upsetting the balance of the others to some extent. A common procedure is to replace the element being tested by an equivalent amount of sodium if it is a cation, and by chloride if it is an anion, but this is not entirely satisfactory as high concentrations of sodium and chloride sometimes inhibit growth.

Using the technique of solution culture, nineteenth-century investigators, including Sachs, Knop and Pfeffer, established that *seven* elements in addition to carbon, hydrogen and oxygen, are required for normal growth. These 'mineral' elements are *nitrogen* (N), *phosphorus* (P), *sulphur* (S), *potassium* (K), *calcium* (Ca), *magnesium* (Mg) and *iron* (Fe). This list differs in only one respect from that compiled in 1804 by the Genevan botanist, Theodore de Saussure, from a study of the salt content of plants, namely in the substitution of iron for silicon. The seven elements are required in relatively large amounts and are consequently referred to as 'major' elements or 'macronutrients'. It is evident from Table 2 that the requirement for iron is appreciably lower than for other macronutrients and for this reason some authors (e.g. HEWITT, 1963) include it among the micronutrients (see below). If any one of these seven elements is omitted from the medium characteristic symptoms of malnutrition rapidly become evident (see below), the plants grow less quickly and eventually die.

Since the last decade of the nineteenth century, further research in which more highly purified chemicals and more refined techniques were used, has established that at least *seven* additional elements are required by plants. These are copper (Cu), manganese (Mn), zinc (Zn), sodium (Na), boron (B), molybdenum (Mo) and chlorine (Cl). They are required

(a)

Plate 1(a) Calcium deficient tomato fruits showing 'blossom end rot'.

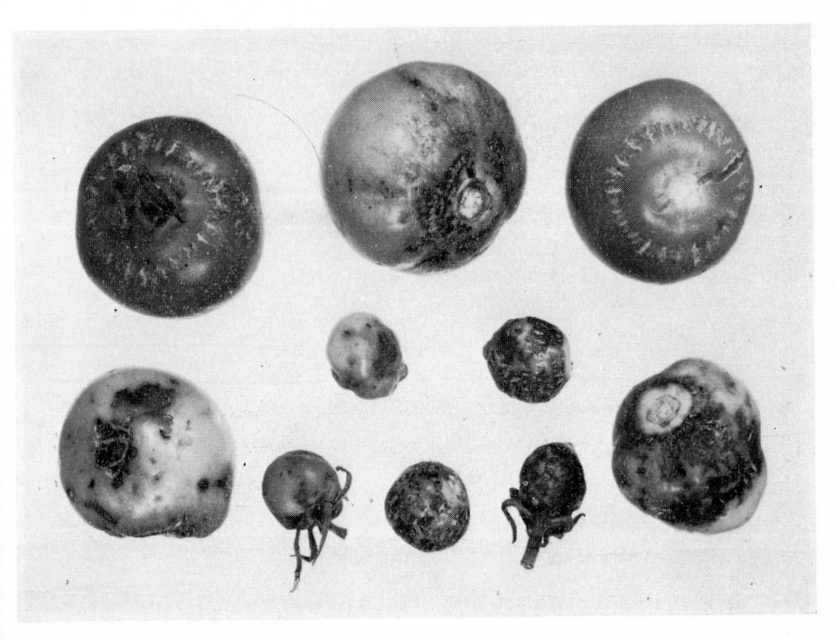
(b)

Plate 1(b) Boron deficiency in tomato: small malformed fruit with corky lesions especially in a ring round the calyx. (Photographs by courtesy of E. J. HEWITT and Long Ashton Research Station, Bristol)

(a)

Plate 2(a) Iron deficiency in tomato. Note interveinal chlorosis more pronounced in basal and central regions of leaflets; no necrosis.

Plate 2(b) Manganese deficiency in tomato. Orange-yellow chlorotic mottling with early development of profuse brown necrotic spots. (Photographs by courtesy of E. J. HEWITT and Long Ashton Research Station, Bristol)

at relatively low concentrations compared even with iron and so are called 'trace', 'minor elements' or 'micronutrients'. Despite the small amounts required they are no less essential than the macronutrients because plants cannot survive without them. If there is an insufficient supply of any micronutrient, deficiency symptoms become visible and the plants die, sooner or later. Some of the symptoms of deficiency 'diseases' were noticed and named before the underlying cause was understood. Thus 'heart-rot' of sugar beet (boron deficiency), 'reclamation disease' (copper deficiency), 'whip-tail' of cauliflowers (molybdenum deficiency) and 'little leaf' of apples (zinc deficiency) are caused by shortage of specific micronutrients and can be cured only by supplying the appropriate element.

The list of micronutrients is undoubtedly incomplete. The difficulty in establishing a definitive list arises from the small amounts of these elements which are required and the impossibility of excluding them entirely from the plant or its surroundings. In the first place a supply of essential elements is carried over from generation to generation in the seed (Table 3). Furthermore, there is a limit to the degree of purification of water and chemicals that can be achieved, and there is always some contamination from culture vessels and from the air. Because of such problems it was not until 1954 that chlorine (as chloride) was shown conclusively to be an essential element for higher plants and it has still not been confirmed for more than a few species. For similar reasons, sodium has only recently been proved to be an essential element for any higher plants other than marine species. As techniques improve it is likely that the list of essential elements will be extended, perhaps to a quite considerable extent. The known essential elements have an interesting distribution within the periodic table. All, except molybdenum, are among the 30 lightest elements and they are clustered in such a way that with the exception of hydrogen every essential lies adjacent to at least one other either horizontally, vertically or (in the case of molybdenum) diagonally.

2.3 Nitrogen, sulphur and phosphorus

2.3.1. Occurrence in soil

These three elements are present in soils in both inorganic and organic substances. *Nitrogen* occurs as nitrate, nitrite, ammonium and free ammonia dissolved in the soil solution. Most of the nitrogen in the soil, however, occurs in organic form partly in nitrogenous plant and animal remains (humus), partly in micro-organisms, and partly in soluble organic compounds, e.g. amino-acids and amides. As rocks contain little or no nitrogen, replenishment of the soil solution depends on breakdown of humus and nitrogen fixation. Gaseous nitrogen in the air spaces of the soil is unavailable to many plants, but micro-organisms can fix it to form nitrogen compounds such as nitrate and ammonia. (Fig. 1–2, p. 2).

Humus also contains *sulphur* and *phosphorus* as constituents of various organic molecules from which they are released, mainly as sulphates and phosphates, through the activity of micro-organisms. The bulk of the sulphur in fertile soils is present as insoluble minerals such as pyrite (FeS_2), sphalerite (ZnS), chalcopyrite ($CuFeS_2$) and gypsum, which are

Table 3 Amounts of some essential elements in tomato seeds from plants grown in a complete nutrient medium and in media deficient in one element for one generation (from BOWEN, 1966)

Element	*p.p.m. of dry matter* complete medium	*p.p.m. of dry matter* deficient media
Boron	9.6	6.8
Calcium	650	360
Chlorine	70	25
Copper	6.8	1.8
Iron	610	23
Potassium	6700	4700
Magnesium	3600	2800
Manganese	55	12
Molybdenum	2.5	0.09
Nitrogen	51 600	44 200
Phosphorus	7200	2000
Sulphur	600	390
Zinc	56	21

slowly oxidized to sulphates, while in arid and semi-arid soils much of it occurs as soluble sulphates. In acid soils most of the inorganic phosphorus occurs as relatively insoluble iron and aluminium phosphates and in association with clay minerals.

2.3.2 *Available forms*

Most higher plants absorb these macronutrients from the soil solution mainly in the oxidized forms (NO_3^-, SO_4^{2-}, $H_2PO_4^-$). Reduced nitrogen (NH_3, NH_4^+) is also taken up quite readily, as are soluble organic nitrogen compounds such as urea ($CO(NH_2)_2$) and amino-acids. Sulphur is absorbed to some extent as the less highly oxidized sulphite (SO_3^-) and thiosulphate ($S_2O_3^{2-}$) ions while the more highly oxidized bi- and tri-valent phosphates (HPO_4^{2-}, PO_4^{3-}) are absorbed less readily than the univalent form. The phosphate concentration in a typical soil solution is usually low in comparison with other macronutrients (Fig. 2–1) but the solution is replenished continuously from phosphorus-containing compounds in the solid phase.

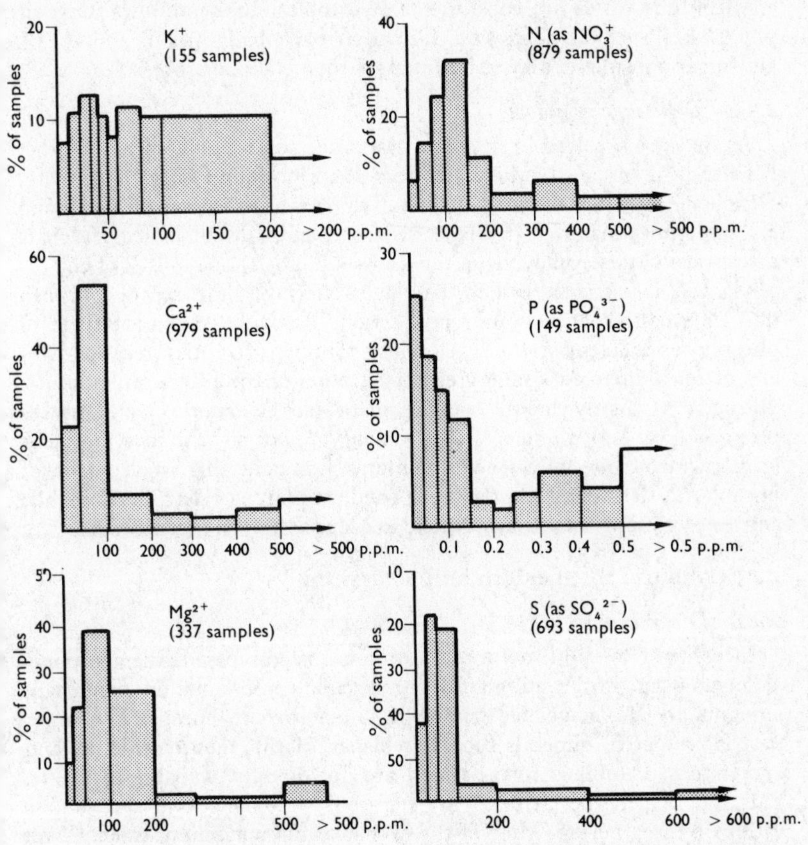

Fig. 2—1 The composition of soil solutions. The height of each rectangle indicates the percentage of samples which fell within the concentration range limited by the vertical bars (after REISENAUER, 1966).

2.3.3 *Functions*

Nitrogen, sulphur and phosphorus are constituents of a variety of organic compounds which are essential to the structure and metabolism of plants. Nitrogen occurs, for example, in nucleic acids, proteins, chlorophyll, and various coenzymes, including the nicotinamide adenine dinucleotides (NAD and NADP). Sulphur is found in some other coenzymes e.g. biotin, thiamine and coenzyme A and in the amino-acids, cystine, cysteine and methionine which occur in proteins. Sulphur bridges (–S–S–) have an important role in determining protein structure and sulphydryl groups (–SH) are often part of the active centre of enzymes. Phosphorus is a component of nucleic acids and also of phospho-lipids, e.g. lecithins which are constituents of cytoplasmic membranes. Organic phosphates such as adenosine di- and triphosphates (ADP and ATP), and sugar phosphates play an essential role in metabolic processes.

2.3.4 *Deficiency symptoms*

As they are involved in so many vital processes, it is not surprising that deficiency of any one of these elements has a profound effect on growth. Deficient plants are weak and stunted while the leaves remain small and are sometimes distorted in shape. Nitrogen and sulphur deficiency characteristically cause yellowing of the leaves (chlorosis) due to lack of chlorophyll and if nitrogen is deficient stems may become red or purplish due to excessive anthocyanin production. Particularly characteristic of phosphorus deficiency is the development of brown (necrotic) areas on leaves and petioles, and a dark blue-green colouration of the leaves often occurs. Nitrogen is highly mobile in the plant (see Chapter 5) and moves progressively into younger leaves during growth so that it is the older leaves which show deficiency symptoms first and this is also true of phosphorus deficiency. On the other hand, sulphur is relatively immobile and so symptoms of deficiency tend to occur first in younger leaves.

2.4 Calcium, magnesium and potassium

2.4.1 *Occurrence in soil*

In contrast to the three macronutrients just discussed, these cationic elements occur in the soil mainly in inorganic compounds but significant amounts are also associated with organic materials in humus. The larger part is present in minerals such as feldspar, biotite, montmorillonite and anorthite in combined forms which are not directly available to plants, but from which the elements are released as exchangeable and soluble cations by weathering. Most of the exchangeable potassium, calcium and magnesium is associated with clay particles. Depletion of the soil solution of soluble cations by plant roots and micro-organisms causes a further release of soluble cations from the exchangeable fraction of the solid phase which in turn is replenished from the non-exchangeable fraction. The cation exchange capacity (C.E.C.) of a soil is a measure of the amount of exchangeable cations present and can be determined by treating a soil with dilute acid which causes replacement of the metallic cations by hydrogen ions. Soils derived from calcareous rocks (e.g. chalk and limestone) are rich in calcium and also tend to have higher values of C.E.C. than those derived from silicious rocks.

2.4.2 *Available forms*

Potassium, calcium and magnesium are absorbed mainly as cations from the soil solution. The majority of soils contain these elements at less than 200 p.p.m. (Fig. 2–1) but occasionally values in excess of 1000 p.p.m. have been found. There is some evidence that uptake may also take place directly from the solid phase by 'contact' or 'proximity' exchange (see p. 40). Increasing acidity of the soil solution tends to decrease the availability of these cations (Fig. 2–2) because a high proportion of the cation exchange sites in an acid soil are occupied by hydrogen ions. Hydrogen ions also compete with metallic cations for carrier sites in the root (see p. 38).

2.4.3 *Functions*

Magnesium is an essential constituent of chlorophyll and is also associated with many plant proteins. Partial removal of magnesium from ribosomes results in a highly characteristic dissociation of the particles into sub-units. Magnesium ions are the natural activators of a number of enzymes including nearly all of those acting on phosphorylated substrates.

Calcium, in the form of calcium pectate, is an important component of plant cell walls. Calcium salts of phosphatidic acid occur in membranes and are essential to the maintenance of their structure and properties. Amylase is activated specifically by calcium while a number of other enzymes which require magnesium such as ATP-ases are activated, but to a lesser extent, by calcium. The presence of large amounts of insoluble calcium salts of organic acids, e.g. oxalic acid, in many plants suggests that it may have a role in regulating the acidity of cell sap; alternatively it is possible that the acid is produced to precipitate calcium ions rather than the reverse. Fungi require much less calcium than do green plants and for them calcium can be considered a micronutrient. They accumulate organic acids mainly as soluble potassium salts.

Potassium is not known to be a constituent of any essential organic molecule other than those acids which form potassium salts. It forms loose associations with proteins and is an activator of pyruvate kinase and numerous other enzymes. Over forty enzymes are known which require a univalent cation for maximum activity; potassium is usually the most effective, although rubidium and ammonium can also function, at least *in vitro*; sodium and lithium ions are often inhibitory. Transport ATP-ases (see p. 36) from animal sources require sodium as well as potassium ions for maximum activity. No such synergistic effect of the two cations has been found in plants although often plant ATP-ases are activated *in vitro* by sodium as well as by potassium.

If the function of potassium is only to activate enzymes it is at first sight surprising that it is required by plants in such large amounts (cf. micronutrients, see below). An explanation of this apparent anomaly may be that because potassium has a very low affinity for protein a high concentration is needed to maintain the potassium-protein complexes essential for optimal enzyme activity. For maximum activity of potassium-requiring enzymes, such as pyruvate kinase, *in vitro*, concentrations as high as 100 mM seem to be required. However, this does not explain why most of the potassium taken up by plants is accumulated in vacuoles. If this did not occur and potassium ions were concentrated solely in the cytoplasm the potassium requirement of plants would presumably be much lower. Clearly, potassium has an important role as an osmotic regulator and for some reason it cannot usually be satisfactorily replaced in this respect by another cation although sodium does substitute for it to a considerable extent in halophytes. Plants seem to have an almost unlimited capacity to accumulate potassium but above a certain level growth is not improved by additional absorption ('luxury' consumption). It has recently been discovered that potassium salts, mainly of organic acids, accumulate in guard cells when stomata open and are released when they close. This contributes to a change in osmotic (solute) potential upon which the turgor of the guard cells depends. When leaves wilt there is an increase in the concentration of an inhibitor, abscisic acid, and this seems to cause the stomata to close, apparently by facilitating release of potassium ions from the guard cells.

2.4.4 *Deficiency symptoms*

In the absence of potassium, growth is much reduced especially in plants which have few reserves in the seed (Fig. 1–6). Moderate deficiency which may not affect growth appreciably, typically results in a mottled chlorosis of the leaf followed by development of necrotic patches at the tips and margins of the leaves. These patches have been found to contain large amounts of an unusual nitrogenous compound, putrescine, which is produced as a result of disturbed metabolism and eventually kills the cells. The leaves sometimes develop a metallic sheen (bronzing) prior to the development of other symptoms and later they may curve downwards and roll inwards towards the upper surface. Deficiency symptoms appear first in the older leaves indicating that potassium is mobile in the plant and moves towards younger tissues (cf. nitrogen and phosphorus); the same is true of magnesium. Characteristic symptoms of magnesium deficiency are extensive interveinal chlorosis followed by accumulation of anthocyanin pigment and necrosis. Magnesium, like potassium, usually causes reduction of growth, marked shortening of internodes, premature death of leaves and inhibition of flowering.

Calcium deficiency results in early death of meristematic regions of stem and root; malformation of the young leaves, causing the tips to be hooked back, is also a characteristic symptom. Later, the leaves may show marginal chlorosis and these areas eventually become necrotic. Once it is deposited in leaves, calcium, like sulphur is immobilized (see Chapter 5) and symptoms of deficiencies tend to develop in young leaves as soon as supply is depleted. In the absence of calcium roots do not grow well and often appear brown in colour and stunted. The presence of magnesium appears to enhance this effect. Degeneration at the apex of young fruits ('blossom end rot') is a common symptom of calcium deficiency in tomatoes (Plate 1 (a)).

2.5 Iron

2.5.1 *Availability in soils*

Most soils contain appreciable quantities of iron in the form of hydrated oxides such as limonite ($Fe_2O_3.3H_2O$) and as insoluble sulphides, from which it is slowly released as soluble ions. The ferrous form (Fe^{++}) is readily absorbed by plants and under alkaline conditions it is oxidized to the ferric form (Fe^{+++}) which is relatively unavailable. Thus plants tend to suffer from iron deficiency in well-aerated calcareous soils, which are usually alkaline, unless iron is present in organic complexes which are taken up irrespective of pH.

2.5.2 *Functions*

Iron occurs in the prosthetic group of certain proteins, notably the cytochromes which function in electron transport, and in the enzymes, peroxidase and some dehydrogenases. It has a specific role in chlorophyll synthesis, because an iron-porphyrin compound is formed as an intermediate, but it does not occur in chlorophyll itself. Ferredoxin is an iron-containing protein which acts as an electron carrier in photosynthetic phosphorylation and in nitrogen fixation. Haemoglobin, another haem-protein, is found in nitrogen-fixing root nodules.

2.5.3 *Deficiency symptoms*

Iron deficiency (Plate 2(a)) usually leads to extensive chlorosis of the leaves. The veins of the leaf remain green longest, probably because the concentration of iron is highest there. Younger leaves are most affected because iron does not readily move out of the older leaves (cf. sulphur and calcium). There is an abrupt cessation of cell division in the apical meristem of excised pea roots after about 7 days' growth in a medium containing no iron, and iron deficiency also inhibits the production of leaf primordia in shoot apices. The cause of this inhibition of cell division is still not understood. An excess of certain metal cations, such as manganese, copper, zinc or nickel may induce symptoms similar to those of iron deficiency (see p. 21), possibly because of competition for iron acceptor sites within the plant.

2.6 Copper, manganese and zinc

2.6.1 *Occurrence in soils*

The major copper compound in primary rocks is chalco-pyrite ($CuFeS_2$) from which natural deposits of copper sulphide have probably originated. From both of these sources copper is gradually released by weathering. Bivalent copper ions are strongly adsorbed on to clay particles in an exchangeable form and copper also forms stable complexes with organic molecules. The concentration of free copper ions in soil solutions is usually low; concentrations above about 1 p.p.m. are toxic to many micro-organisms and cause a decrease in soil fertility.

Manganese occurs as insoluble oxides in soil and, like copper, exists in exchangeable forms associated with soil colloids and as organic complexes. The most important zinc-bearing mineral is the sulphide, sphalerite, found in igneous rocks. Zinc is also present in ferro-magnesium minerals such as magnetite, biotite and hornblende which are readily decomposed to release bivalent zinc, some of which is absorbed by clay particles or complexed with organic matter.

2.6.2 *Available forms*

These elements are absorbed most readily as bivalent ions (Cu^{2+}, Mn^{2+} and Zn^{2+}). In well-aerated soils and at high pH, Mn^{2+} is oxidized to tri- or tetravalent ions which are absorbed less readily and for this reason sufficient manganese is often not available in calcareous soils (Fig. 2–2). Copper and zinc availability also decreases with increasing pH,

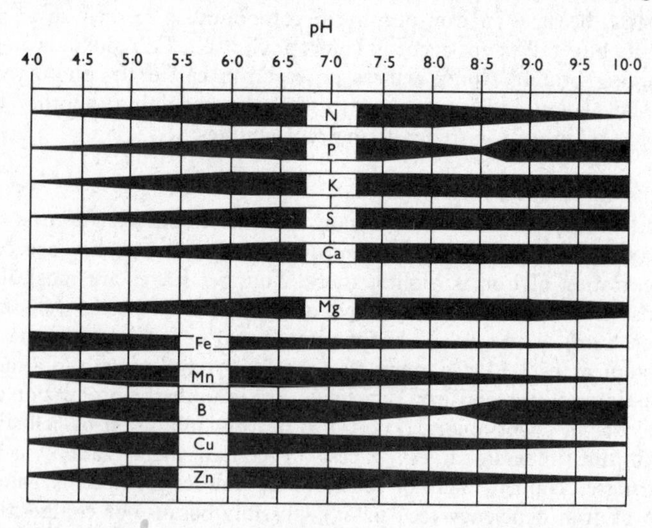

Fig. 2–2 General relationships between soil pH and availability of individual plant nutrient elements. (Redrawn from TRUOG, 1951)

and with increasing levels of phosphate because of precipitation as insoluble copper and zinc phosphates. In contrast to iron and magnanese, copper and zinc are not absorbed readily as organic complexes and this sometimes accounts for copper deficiency in humus-rich soils, such as those of the Everglades in Florida.

2.6.3 *Functions*

Copper is a component of several metallo-enzymes including ascorbic acid oxidase, phenolase and cytochrome oxidase. The copper appears to act as an intermediate electron acceptor in the direct oxidation of substrates by molecular oxygen through its ability to undergo reversible oxidation and reduction between the cupric (Cu^{2+}) and cuprous (Cu^{+}) forms. Manganese, like magnesium, activates a number of phosphate-transferring enzymes. The relative importance of manganese and magnesium in enzyme activation *in vivo* is not quite clear; presumably there are some enzymes which have a specific requirement for manganese since it cannot be entirely replaced by magnesium, or any other element. A mangano-protein, manganin, has been isolated from pea-nuts but its role in metabolism is uncertain.

2.6.4 *Deficiency symptoms*

Copper deficiency is well known in citrus and other deciduous fruit trees in many parts of the world. Leaves become dark green and twisted and later develop interveinal mottling and necrosis. The main shoots may die back and lateral buds grow out giving a 'witches broom' effect. The bark becomes roughened and eventually splits, causing gum to exude, a conditon known as *exanthema*. In cereal crops, copper deficiency causes the young leaves to become limp and chlorotic and remain tightly rolled while the tip turns white and eventually collapses ('white tip'). In tomato plants the foliage is at first abnormally dark blue-green, then becomes paler and inrolled before premature withering (Plate 3).

The visible effects of manganese deficiency are more diverse than for any other mineral element. Symptoms may appear first on either young or old leaves and comprise a wide variety of chlorotic and necrotic patterns (Plate 2(b)). Some plants including apples show interveinal chlorosis which stands out against a net-work of green veins (cf. iron deficiency). In potatoes (*Solanum tuberosum*), chlorosis is preceded by the appearance of dark necrotic spots close to the mid-rib of the leaf. 'Grey speck' of oats (*Avena sativa*) is so named because manganese deficiency causes necrotic areas to appear which dry out with a pale grey-green halo round each of them.

A failure of leaves to expand and stems to elongate is a characteristic symptom of zinc deficiency, causing the conditions known as 'little leaf' and 'rosette leaf' in apples, peach (*Prunus persica*) and pecans and tomato (Plate 3). These effects are believed to be associated with a disturbance of auxin metabolism. Zinc deficient plants have a low level of indol-3yl acetic acid (IAA) and this has been attributed to a requirement for zinc in the synthesis of tryptophan which may be a precursor of auxins.

2.7 Molybdenum and boron

Molybdenum and boron occur in soil solutions mainly as molybdates ($HMoO_4^-$ and MoO_4^{2-}) and boric acid (H_3BO_3) a covalent compound which by loss of water gives rise to the borate ion (BO_2^-). The concentration of these elements in the soil solution is exceedingly low. The total amount of molybdenum in soil is rarely more than about 5 p.p.m. of which 90% may be insoluble. It is adsorbed on to soil particles in an exchangeable form but occurs mainly in soil minerals such as molybdenate (MoS_2), powelite ($CaMoO_4$) and wulfenite ($PbMoO_4$) in non-exchangeable forms. Exchangeable boron occurs as insoluble borates, especially of calcium and magnesium. Boron is also present in tourmaline, a very insoluble fluoroborosilicate and in iron and aluminium complexes.

2.7.1 *Availability*

Molybdenum is taken up most readily as the bivalent ion and thus in contrast to other micronutrients uptake is favoured by high pH (Fig. 2–2). On the other hand, raising the pH causes boron to become less available and so absorption is reduced by addition of lime to soils. Differences in the solubilities of calcium, potassium and sodium metaborate may contribute to this effect. The available boron in a soil can be determined by extraction with boiling water which removes most of the substances which can supply boron to plants.

2.7.2 *Functions*

Molybdenum is essential for nitrogen fixation and for nitrate assimilation. It has already been noted (p. 11) that those organisms which utilize atmospheric nitrogen have a considerably higher molybdenum requirement than those which do not. Molybdenum is believed to serve as an activator of enzymes involved in these processes. HEWITT (1963) reported that molybdenum deficiency causes an accumulation of inorganic phosphate in tomato and cauliflower (*Brassica oleracea*) leaves and it is possible therefore that it is also involved in some phosphorylation reactions, which perhaps involve the formation of organic phospho-molybdate complexes.

Boron is the one micronutrient which has not been shown to act as a specific activator or metal component of any enzyme. Examination of the effects of boron on various enzymes indicates that it may act as an inhibitor rather than an activator and it has been suggested that it functions in this way as a regulator of metabolism. The borate ion complexes with various polyhydroxy compounds including some sugars and GAUCH and DUGGAR (1953) suggested that a major role of boron is in the translocation of sugars across cell membranes. However, the observation that translocation of sucrose is reduced in boron-deficient tomato plants may be attributed to an inhibition of utilization of materials in the meristems rather than to a direct effect on transport. Boron has been found to stimulate germination in a grass (*Themeda triandra*) and this may be due to an effect on synthesis of gibberellic acid.

2.7.3 *Deficiency symptoms*

Molybdenum deficiency in tomatoes grown on nitrate leads to chlorotic interveinal mottling of the lower leaves followed by marginal necrosis and infolding. 'Whiptail' in brassicas is a characteristic indicator of molybdenum deficiency. The leaf blade withers leaving only the mid rib and adjoining leaf tissue giving the appearance of a whip. Flower formation is decreased or suppressed by molybdenum deficiency and if flowers do form they absciss before setting fruit.

Boron deficiency causes disorganization of meristems and early death of stem tips in many plants. Leaves may become crinkled and mis-shapen and petioles and stems crack. Storage organs are sometimes affected as in 'heart rot' of sugar beet (*Beta saccharina*) and 'water core' of turnips (*Brassica rapa*). Flowering is often totally suppressed while fruit and seed formation, if it occurs, is abnormal. Apples and tomatoes are particularly susceptible to boron deficiency and the fruits become malformed, with a corky surface and hard core (Plate 1(b)).

2.8 Chlorine and sodium

2.8.1 *Availability in soils*

These elements occur in the soil solution as univalent ions and are sometimes present at high concentration, e.g. in soils near the sea or salt lakes. Sodium becomes absorbed by clay colloids and at high concentration causes displacement of potassium and calcium, leading to deterioration of soil texture. Soluble and exchangeable sodium and chloride are readily leached from soils and land reclaimed from the sea, e.g. in Holland, becomes normal in 3–5 years.

2.8.2 *Functions*

The role of these elements in plant metabolism is still uncertain. The observation that chloride is essential for production of oxygen by isolated chloroplasts has led to the view that chloride acts as an electron-transporting agent in photophosphorylation. Sodium is an activator of transport ATP-ases in animals and possibly also in plants. There is evidence that sodium can replace potassium partly in some of its functions.

2.8.3 *Deficiency symptoms*

Because sodium and chloride are so ubiquitous in nature and such small amounts are evidently required by most plants, deficiency symptoms have hardly ever been observed although the growth of many plants is reduced in soils low in common salt. Growth of lettuce, tomatoes, cabbage and carrots (*Daucus carota*) is reduced by more than 50% in chloride-deficient media. Chloride deficiency symptoms have been induced in tomato plants; they include reduced growth and some wilting which is followed by chlorosis, bronzing and necrosis. Root growth is also markedly affected; the roots become stunted and development of laterals is suppressed.

Addition of sodium salts, especially sodium chloride, to soil stimulates the growth of some plants, notably sugar beet, red beet, celery and turnips and sometimes induces 'succulence' but it severely inhibits the growth of others. The effects of sodium are particularly evident when potassium is deficient. There is now good evidence that sodium is an essential micronutrient for *Atriplex vesicaria* and some other plants, notably those showing the C_4 photosynthetic pathway (BROWNELL and CROSSLAND, 1972).

2.9 Beneficial elements

Addition of a particular element to the soil or culture solution may lead to improved growth without that element being necessarily essential. Rubidium, strontium, cobalt, selenium, aluminium and silicon are among the elements which have been shown to have such beneficial effects. In some cases, e.g. rubidium and strontium, the element may substitute in part for an essential element, in this case potassium or calcium, and thus produce beneficial effects when this element is deficient. Alternatively, the beneficial element may stimulate absorption or transport of an essential element which is in limited supply or conversely inhibit uptake and distribution of one that is in excess (2.10).

Growth stimulation may also be observed in cases where there is no apparent discrepancy of an essential element. This may occur if the beneficial element is more efficient in one particular function than an essential element without being able to substitute in other functions, or if it stimulates a metabolic process which facilitates growth, but is not absolutely vital.

2.10 Toxic effects of mineral elements

Growth is impaired if the concentration in the medium of either essential or non-essential elements exceeds a certain level. In general, macronutrients are much less toxic than micronutrients and their concentration can be raised appreciably above the optimum without significantly affecting growth. This is particularly true of potassium, and most plants take up this element under favourable conditions in excess of their apparent needs without ill-effects (see p. 18). On the other hand, the margin between sufficiency and toxicity is very narrow for most micronutrients, and notably for boron. However, chlorine is unusual in this respect as many plants can tolerate relatively high concentration even though their chlorine requirement is small. Susceptibility varies greatly between species and can sometimes be related to the ease with which the element is absorbed. Manganese uptake, for example, is low in sugar beet and oats, which are relatively tolerant species compared with the kidney bean (*Phaseolus vulgaris*) and cabbage, which are susceptible to manganese toxicity. In other cases, tolerance appears to depend on an ability to immobilize toxic ions, e.g. in cell walls, or in vacuoles.

A similar situation exists for non-essential as for essential elements and whereas some, e.g. silicon, are innocuous, others, notably arsenic, chromium and silver, are usually highly toxic even at low concentrations. It is the toxicity of such elements that inhibits the colonization by many plants of slag-heaps and other industrial waste.

An excess of a particular element may injure plants in several ways. Toxicity may result from a low water potential, as in saline soils, and in addition there may be specific effects of individual elements. A high concentration of one element may lead to deficiency of another by interfering with uptake in which case the toxicity symptoms of one element may resemble those for deficiency of another. For example, an excess of aluminium produces symptoms resembling phosphate deficiency because it causes precipitation of phosphate as insoluble aluminium phosphate in the soil and in the root tissue. Why the so-called 'aluminium accumulator' plants such as *Hydrangea macrophylla* can tolerate amounts of aluminium in their tissues sometimes in excess of 40 000 p.p.m. is not yet clear.

Excess manganese, copper, zinc, cobalt and several other metallic ions induce symptoms resembling iron deficiency in some plants. The relative activity of the various metals is in the order of stability of the chelation complexes they form with organic molecules, and so metallo-organic complex formation has been suggested as the probable explanation of metal-induced iron deficiency. TWYMAN (1951) postulated that the effect of manganese in causing iron deficiency is due to competition between manganese and iron for an iron-acceptor site in the plant. Excess manganese does not usually reduce the level of iron in leaves showing iron deficiency which suggests that iron is merely displaced from its site of action.

The sensitivity of different parts of the plant to high concentrations of individual elements varies greatly. It has been observed, for example, that the level of boron which produces optimal growth in young sunflower leaves is toxic to older leaves. This may account for the fact that whereas boron deficiency symptoms usually appear first near the tip of the shoot (p. 23), toxicity symptoms develop initially in older leaves.

Concentrations of a particular element which are toxic to one species may be beneficial to another. Calcium reduces the uptake of manganese, iron, boron and zinc and this may account for the apparent toxicity of calcium to calcifuge plants (see p. 7). On the other hand, high levels of calcium may promote the growth of some calcicoles because in its absence the concentrations of other elements in the tissues reach toxic levels.

3 Ion Absorption by Cells

3.1 Membrane potential

An ion in aqueous solution is acted on by at least two physical forces—one arising from chemical potential gradients and the other from electrical potential differences. Chemical potential is related to the concentration of the ion and electrical potential is the result of the net positive or negative charge carried by the ion. When a salt is added to water it diffuses through the solution from regions of higher concentration to those of lower concentration until a uniform concentration is achieved. In general, one of the ion species will have a higher mobility than the other and will tend to diffuse more quickly than its oppositely-charged partner and thus cause a slight separation of charges. This sets up an electrical potential gradient leading to a *diffusion potential.* As a result the faster moving ion of the pair is slowed down and the slower one speeded up until they both move at the same rate. Thus a dissociated salt diffusing in a solution behaves as a single substance and has a characteristic *diffusion coefficient.* According to Fick's law, the rate of diffusion dv/dt is related to the concentration gradient dc/dx thus:

$$dv/dt = -Da\, dc/dx \qquad (3.1)$$

where D is the diffusion coefficient and a the area across which diffusion occurs. The negative sign is a convention to indicate that diffusion occurs from a higher to a lower concentration.

The cell surface membrane or plasmalemma is usually the main barrier for the diffusion of molecules into and out of cells. Diffusion of ions across such a membrane results in the development of a diffusion potential across it which is termed the *membrane diffusion potential.*

The relationship between the electrical potential difference across a membrane and the accompanying distribution of an ion across it at equilibrium is given by the following equation:

$$E_j = \frac{RT}{z_j F} \ln \frac{C_o}{C_i} \qquad (3.2)$$

where E_j is the electrical potential between inside and outside for ion species j; R is the gas constant; T, the absolute temperature; z_j, the valency of the ion; F, the Faraday; ln is $\log_e$, and C_o and C_i, the external and internal concentrations of ion species, j. E_j is called the *Nernst potential* of ion species j and (3.2) is the Nernst equation. When C_o and C_i are expressed in millimoles litre^{-1}, E_j is in millivolts.

The Nernst equation can be simplified to the form (at 18° C)

$$E_j = 58 \log_{10} \frac{C_o}{C_i} \text{ for cations} \tag{3.3}$$

and

$$E_j = 58 \log_{10} \frac{C_i}{C_o} \text{ for anions} \tag{3.4}$$

Thus if, for example, potassium ions were present at 1 mM outside and 10 mM inside then, from the Nernst equation, an electrical potential of −58 mV (−ve inside with respect to outside) would maintain that concentration gradient across a membrane in a passive equilibrium state. On the other hand the maintenance of the same concentration gradient for chloride ions would require a potential difference of +58 mV. Thus a relatively small electrical potential difference can balance a large difference in concentration across a membrane. If the electrical potential across the cell membrane was found to be different from the −58 mV predicted by the Nernst equation in the example just given, then it is evident that work must be done to maintain the concentration gradient of potassium in that system. If the measured value was less negative, say −30 mV, then work would have to be done to pump potassium into the cell to maintain potassium at 10 mM inside. If the measured value was more negative, say −90 mV, then work would have to be done to prevent potassium reaching a higher level than 10 mM, that is, the ion must be pumped out.

It is therefore possible, from a knowledge of the concentration gradient for an ion across a cell membrane, to predict the electrical potential gradient required for passive equilibrium, and then by measuring the actual electrical potential with microelectrodes to estimate the minimum amount of energy needed to transport ions across the membrane. This is proportional to the difference between the calculated Nernst potential and the actual measured potential. To illustrate this, in Table 4 are some data obtained for the large internodal cell of the alga *Nitella*. The internal and external concentrations of sodium, potassium and chloride and the Nernst potential for these ions are given. The measured potential across the membrane was −138 mV. The minimum potential difference needed to move a mole of sodium is therefore (−138)−(−67) or −71 mV; for

Table 4 The concentrations and potentials of various ions in *Nitella translucens*. The subscript *o* refers to the concentration in the external bathing solution; *i* refers to the cytoplasm. The Nernst potentials, E_j, were calculated from equation 3.2 (data from SPANSWICK and WILLIAMS, 1964)

Ion	C_o(mM)	C_i (mM)	E_j (mV)
Na^+	1.0	14	−67
K^+	0.1	119	−179
Cl^-	1.3	65	−99

potassium $(-138)-(-179)$ or $+41$ mV; and for chloride $(-138)-(+99)$ or -237 mV. This result implies that sodium is at a higher potential in the external bathing medium, and potassium and chloride are at a higher potential inside the cell. When work has to be done to move an ion across a membrane such movement is termed *active transport* which may be simply defined as the movement of ions against a gradient of electrochemical potential. This implies that energy derived from metabolic processes is involved in moving an ion in the energetically uphill direction across a membrane.

Another type of electrical potential difference encountered in biological systems is that associated with immobile or fixed charges in a solid phase adjacent to an aqueous phase and this is termed the *Donnan potential*. Such a phase boundary occurs in the cell walls of plant cells. The region containing the immobile charged particles is generally referred to as the *Donnan phase*. In cell walls the presence of a large number of immobile carboxyl groups ($R\,COO^-$) associated with pectin and other compounds provides a cation exchange system. The anion exchange capacity, which is usually less than that for cations, is due to the presence of fixed organic cations, such as amines, within the cell wall matrix. Donnan phases also occur in the cytoplasm where the immobile charges are mainly due to proteins. These proteins are fixed in the sense that they are unable to diffuse across either the plasmalemma or the tonoplast.

When the immobile or fixed ions are arranged in a layer the mobile ions of opposite sign occur in an adjacent layer, the two together being referred to as an electrical double layer, such as is shown in Fig. 3–1. From a knowledge of the ratio of the concentrations of any mobile ion

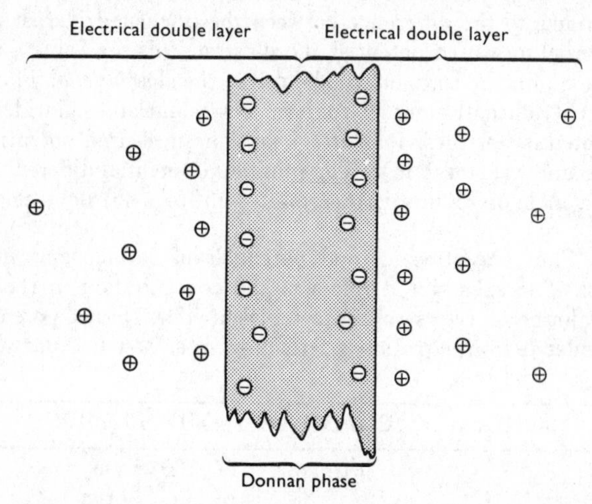

Fig. 3–1 The distribution of positively charged ions, ⊕, occurring on either side of a Donnan phase containing immobile negative charges, ⊖

Plate 3 Left. Zinc deficient tomato plant showing shortened internodes, reduced leaves, leaf curl and epinastic curvature of petioles.
Right. Copper deficient tomato plant showing reduced growth and inrolled leaf margins.
Centre. Control plant of the same age.

(Photographs by courtesy of E. J. HEWITT and Long Ashton Research Station, Bristol)

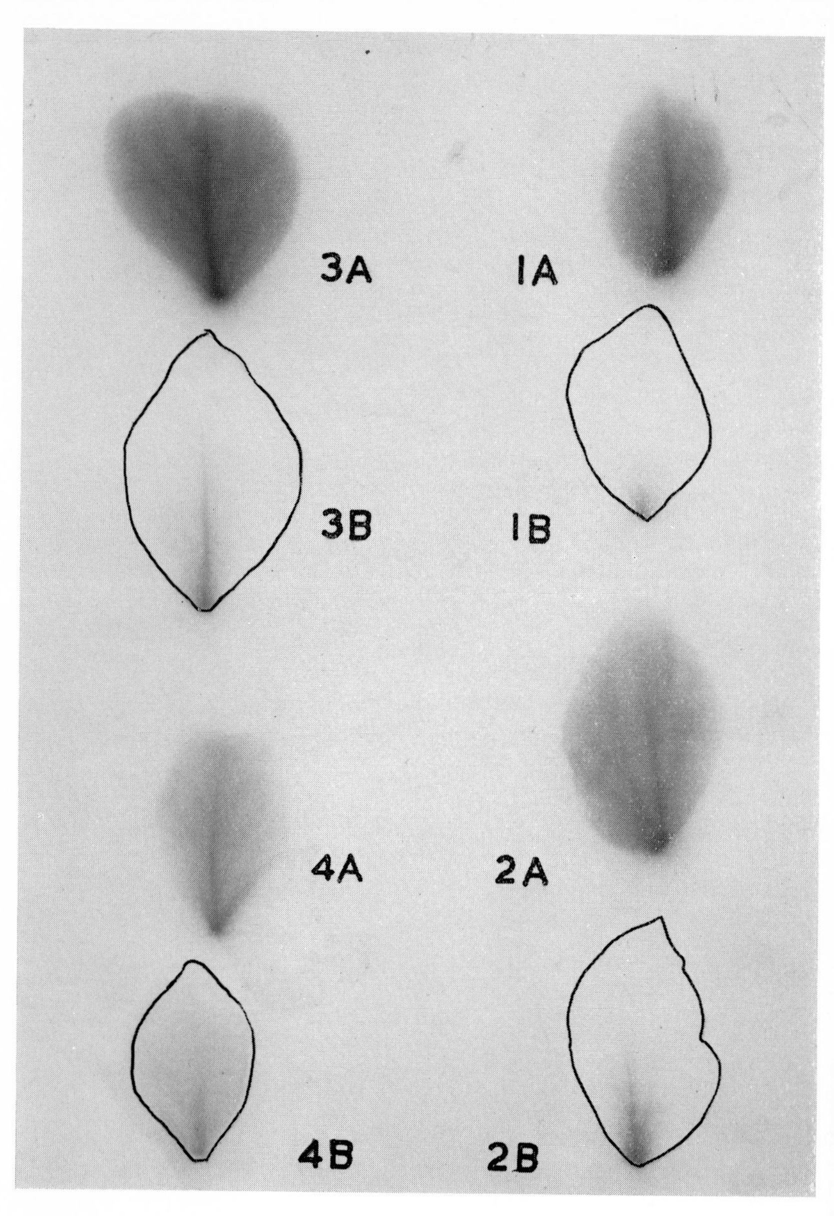

Plate 4 Radioautographs showing the effect of transpiration on movement of rubidium into the leaves of an intact bean (*Vicia faba*) plant. The root system was immersed for 8 hours in a nutrient solution containing ^{86}Rb. The leaves are numbered from the base of the plant upwards and in each case, one of the paired leaflets (A) was allowed to transpire while in the other (B) transpiration was reduced by placing the leaflet in a bag of transparent polythene. (SUTCLIFFE, 1962)

across the electrical double layer the electrical potential difference, or Donnan potential can be calculated, again using the Nernst equation. The Donnan potential may be looked upon as a type of diffusion potential as the mobile ions tend to diffuse away from the charges of opposite sign that are fixed in the Donnan phase.

3.2 Free space uptake

If a plant tissue is first washed in water and then immersed in a solution of a salt there is a rapid initial uptake which is usually completed in 10–20 minutes followed by a less rapid steady uptake which may continue for several hours or even days (Fig. 3–2). The initial rapid uptake is reversible, non-selective and independent of metabolism. If the tissue is transferred back to water a large proportion of the initial uptake may be washed out and this is referred to as the *water-extractable fraction*. A further fraction will be extracted if the tissue is washed in a salt solution with which absorbed ions can exchange. This latter fraction, the *exchangeable fraction*, may be measured more conveniently if the original solution is labelled with a radioisotope (Fig. 3–2).

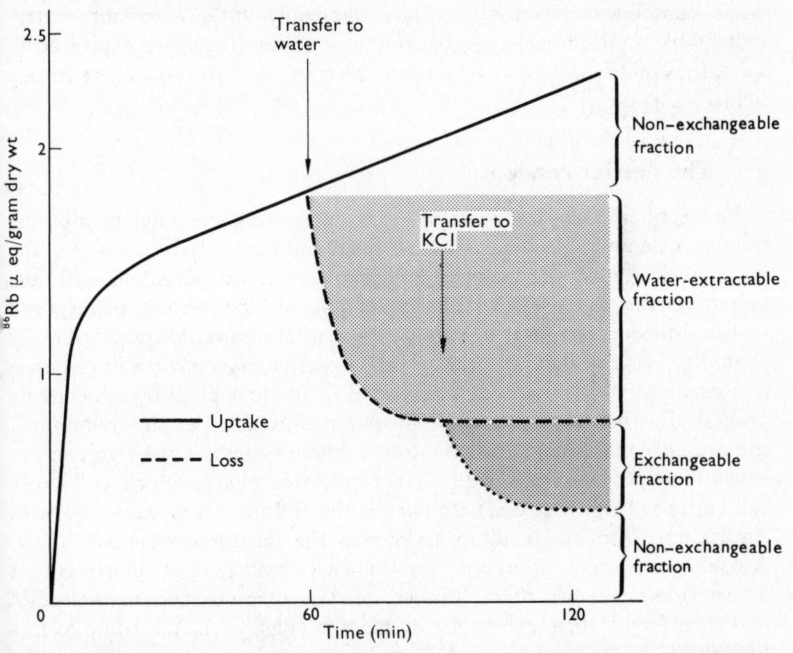

Fig. 3–2 The uptake and loss by maize roots of potassium labelled with ^{86}Rb. The initial rapid uptake may be reversed by washing out the water extractable and the exchangeable fractions with water and unlabelled KCl respectively. The slow steady uptake process evident after 30 minutes is non-exchangeable

It is generally agreed that the rapid initial uptake represents ion movement into water-filled spaces in the cell walls of the tissue and it is usually referred to as uptake into the *free space*, that is into the freely accessible part of the cells. The water-extractable fraction consists of mobile ions in the aqueous phase, or *water-free space*, in the cell wall, while the exchangeable fraction comprises those ions which become adsorbed in the electrical double layer, or *Donnan free space*, which is also in the cell wall.

There have been attempts in the past to estimate the volume of tissue occupied by the free space. This can be done by assuming that the concentration in the free space is the same as that of the bathing medium and converting a quantity of ions taken up per unit volume of tissue to a volume of solution it would occupy. As such an assumption is incorrect, due to the high concentration of ions associated with the electrical double layer, the volumes calculated for the free space in this way are greatly in excess of the actual free space volumes. To avoid this error the concept of *apparent free space* (AFS) was introduced. This may be defined as the calculated volume of a cell or tissue which *would* be occupied if the ion concentration in that volume were the same as that of the bathing medium. This somewhat cumbersome concept is now generally little used and water-extractable and ion-exchangeable fractions of the free space expressed in quantities, not volumes, are the terms used by modern researchers in this field.

3·3 The carrier concept

Ions enter the *non-free space* of a cell by crossing the outer membrane, the plasmalemma. Movement across this membrane may be active, i.e. the ion moves against the gradient of electro-chemical potential, with the help of metabolic energy, or the movement may be passive, in response to the diffusion potential or Donnan potential across the membrane. It should be realized that even the so-called passive movement is in response to a gradient which has been established in the first place by a metabolic process. Furthermore, the synthesis and maintenance of the membrane, the presence of which is essential to the whole system, is itself an energy-consuming process. Ions held in the non-free space, which is mainly cell vacuoles (Fig. 1–4, p. 4), do not readily diffuse out or exchange with similar ions from the medium as long as the membranes remain intact. A useful hypothesis to account for the active transport of solutes is that a membrane constituent or 'carrier' selectively binds certain molecules and then ferries them across the membrane (Fig. 3–3). According to this hypothesis a penetrating ion (M) combines with the carrier (R) at the outer surface of the membrane. This stage might involve adsorption, exchange-adsorption or some kind of chemical reaction. The complex (MR) cannot leave the membrane but is mobile within it and may move to the inner side of the membrane, where it is broken down, releasing the

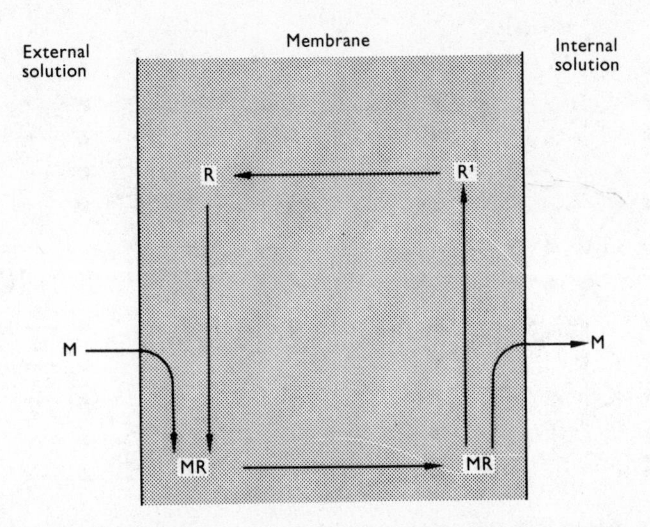

Fig. 3-3 The carrier hypothesis, M is the penetrating ion, R the carrier and MR the mobile complex. R′ is the mobile carrier precursor

ion and forming a carrier precursor (R′). This precursor then moves back across the membrane and is reconverted to R which can now accept another ion at the surface. Thus a limited number of carrier molecules are capable of transporting an indefinite number of ions. An ion of opposite charge to the one actively transported may passively diffuse in response to the potential established across the membrane if the resistance is not too high. If the charge of the actively transported ion is not directly balanced by an accompanying ion of opposite charge then the transport process is *electrogenic*, that is it tends to generate an electrical potential difference across the membrane which will affect the membrane potential.

A widely held view is that ion carrier compounds are proteins and probably enzymes, although positive evidence for this is still awaited. Proteins have a number of properties which are suited to a role in ion transport. They are invariably constituents of biological membranes; they are capable of combining reversibly with specific ions, and they are able to assume various configurations thus altering their shape and position within a membrane from time to time. GOLDACRE and LORCH (1950) proposed that contractile proteins may traverse the membrane, binding ions when in the extended and releasing them when in the contracted condition, thus transporting the ion across the membrane (Fig. 3–4a). Another possibility is that ions may cross membranes by micro-pinocytosis. It is envisaged that ions become bound selectively at specific sites on the membrane and are then transported inwards by an invagination which results in the formation of a micro-vesicle within the cytoplasm (Fig. 3–4b).

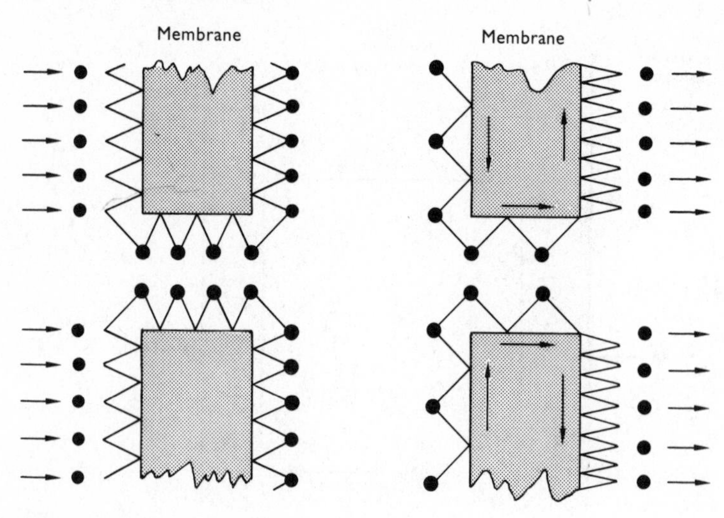

Fig. 3–4 (a) The GOLDACRE hypothesis.

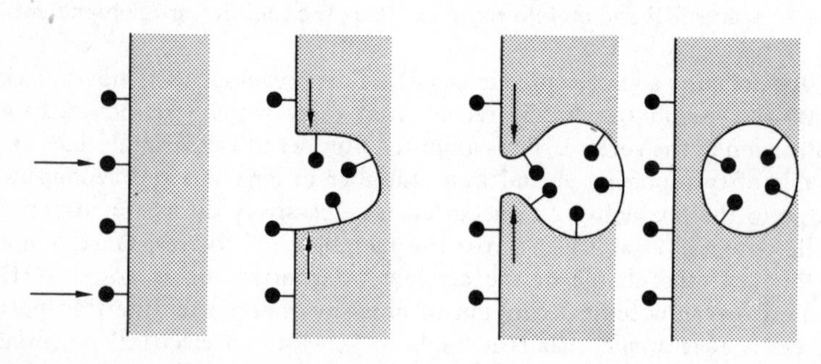

Fig. 3–4 (b) Pinocytosis

Subsequent breakdown of the vesicular membrane could lead to the release of the bound ions within the cytoplasm. Invaginations of the plasmalemma and micro-vacuoles apparently derived from it have been observed in plant cells with the electron microscope and a stimulation of micro-vesicle formation by ions has been demonstrated.

The rate of active ion uptake is often directly proportional to the external concentration of that ion over the lower range of concentrations and then levels off as the external concentration is raised, eventually approaching a maximum or saturation rate (Fig. 3–5a). This relationship is described by the following equation:

$$v = \frac{S \times V_{max}}{K_s + S} \tag{3.5}$$

where v is the rate of active uptake, V_{max} is the maximum rate of uptake when all the carrier is saturated, while K_s is a constant, characteristic of a particular ion crossing a specific membrane, and is expressed in units of concentration (e.g. mM); S is the external concentration of the ion. This equation is analogous to the Michaelis-Menten equation used in analysing the kinetics of enzyme reactions. By analogy with the Michaelis constant, (K_m), K_s is equal to the concentration of solute at which v reaches half maximal rate (Fig. 3–5a). Substituting K_s for S in equation (3.5) we have:

$$v = \frac{K_s \times V_{max}}{K_s + K_s} = \frac{V_{max}}{2} \tag{3.6}$$

When $1/v$ is plotted against $1/S$ a straight line with an intercept of $1/V_{max}$ on the ordinate and a slope of K_s/V_{max} is obtained. This enables K_s and V_{max} for an ion to be calculated accurately. Such plots often yield two lines from which it is possible to obtain two different K_s and V_{max} values (Fig. 3–5b). This result has led to the conclusion that two separate

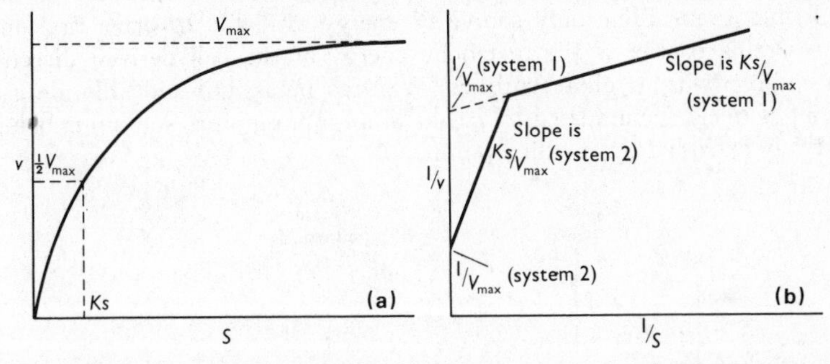

Fig. 3–5 (a) The relationship between the external solute concentration, S, and the rate of active uptake, v, according to a Michaelis-Menten type of kinetics as given by equation (3.5). **(b)** Linear transformation of the Michaelis-Menten equation. This double reciprocal or Lineweaver-Burk plot of ion uptake often yields two lines, as illustrated, from which two different K_s and V_{max} values may be obtained

carriers or carrier sites are involved in the transport of some ion species. One of these sites which appears to be functional at low concentrations (< 1 mM for some cations) is referred to as system 1; the other site, functional at concentrations in the range 1–50 mM, is called system 2. There has been considerable controversy over the location of these two sites, some investigators subscribing to the view that they are in series—system 1 at the plasmalemma and system 2 at the tonoplast (LATIES, 1969) while others believe that the two sites are in parallel and are both at the plasmalemma (EPSTEIN, 1972). Some researchers are sceptical of the concept of dual carriers and think that there may be other interpretations of the data.

Although the carrier hypothesis is the most widely accepted explanation of ion uptake it must be remembered that it is simply a convenient hypothesis which is consistent with the observations that have been made on ion uptake. Before it can be proved that a carrier mechanism operates, it will be necessary to isolate and identify carrier molecules and explain how they work. This has not yet been done although some progress has been made in isolating ion-binding proteins from membranes of micro-organisms.

3·4 The energetics of active transport

In Section 3.1 it was explained how the minimum amount of energy required to move sodium, potassium and chloride ions in *Nitella* can be calculated. The values of the potential difference required, derived from the data of Table 4, −71 mV for sodium, +41 mV for potassium, −237 mV for chloride are a measure of the magnitude of the energy barrier which has to be surmounted by these ions in their movement across the cell membrane. The only source of energy available to drive the ions across this barrier is the metabolic energy of the cell derived directly or indirectly from photosynthesis. Whereas potassium and chloride are being actively accumulated by means of an influx pump, sodium is being actively extruded by an efflux pump.

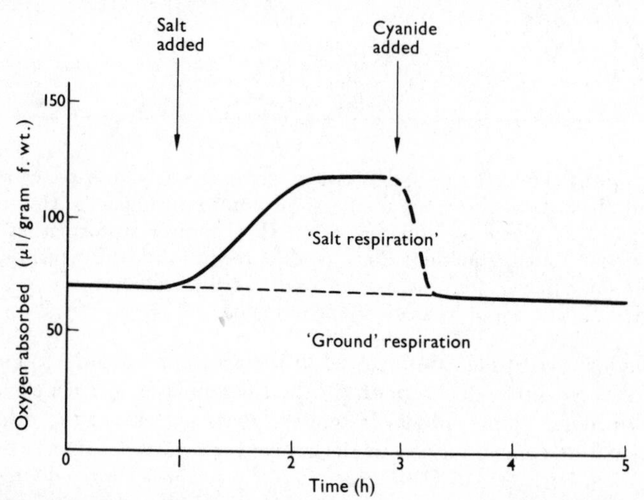

Fig. 3–6 The effect of salt and cyanide on the respiration of plant tissue. The increased respiration above the ground level, the 'salt respiration,' is completely inhibited by low concentrations of cyanide (10^{-5} M). (Based on data from ROBERTSON and TURNER, 1945)

There is some evidence that pumps of this type also occur in higher plants. In non-green tissues, such as roots, the energy required is obtained from respiration. When a tissue which has been washed for some time in distilled water is transferred to a dilute salt solution the respiration rate of that tissue increases (Fig. 3–6). This increased respiration above the ground level has been termed '*anion respiration*' (LUNDEGÅRDH, 1955) and '*salt respiration*' (ROBERTSON, 1968). It is believed that the tissue respires faster to provide extra energy for ion accumulation in response to an increase in adenosine diphosphate (ADP) level in the tissue. The observation that both salt respiration and ion absorption are sensitive to inhibition by cyanide and carbon monoxide, suggests that cytochrome oxidase is implicated somehow in both processes. There is controversy over the question of whether ion transport is energized by ATP directly or whether the process is more directly coupled to electron transfer during respiration. ROBERTSON (1968) believes that the separation of hydrogen ions and electrons in a membrane might drive ion movements, protons exchanging, for example, with potassium ions and hydroxyl ions with chloride. The scheme is presented in Fig. 3–7. The ion transport resulting from this

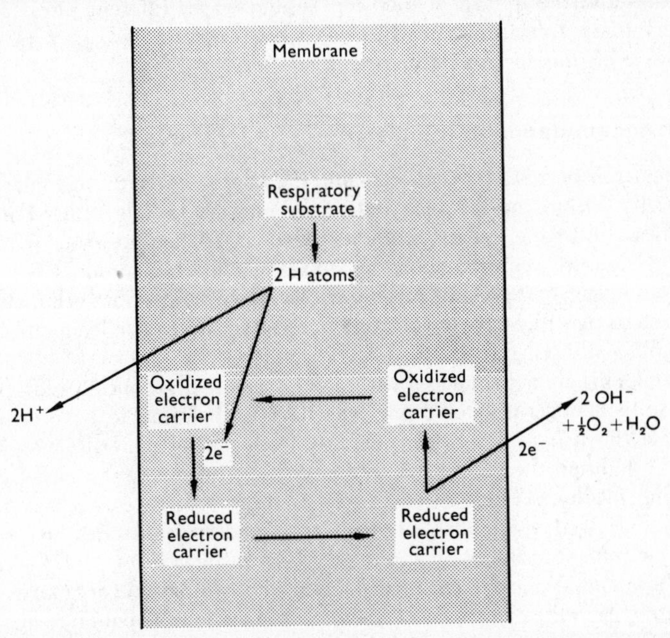

Fig. 3–7 The separation of hydrogen ions and electrons in a membrane impermeable to ions. The charge separation yields hydrogen ions on one side of the membrane while on the other side electrons ejected result in the formation of hydroxyl ions ($2e^- + \frac{1}{2}O_2 + H_2O \rightarrow 2OH^-$). (After ROBERTSON, 1968)

charge separation may be an alternative to the formation of ATP by the same process, as visualized by MITCHELL (see TRIBE and WHITTAKER (1972), in this series). DAINTY (1969) has postulated that in higher plant cells, by analogy with giant algal cells, the anion pump is linked to electron transfer while any cation pumps are energized by ATP direct. The hydrolysis of ATP under biological conditions yields at least 10 Kcal mole $^{-1}$ (1 Kcal $\approx$ 4.2 kT) which is sufficient energy per mole of ATP hydrolysed to transport ions at the maximum rates observed. In addition to operating ion pumps, energy from respiration must also be used to synthesize carriers and binding sites and to maintain membrane structure.

There is also evidence that the ability of a plant cell to accumulate ions is closely related to its capacity to synthesize proteins. If the ion carrier is a protein then ion uptake may be prevented when the synthesis of protein is inhibited, assuming there is a turnover of the carrier. Reversible inhibition of ion uptake has been obtained with some antibiotics which are thought to interfere specifically with protein synthesis. In the presence of cycloheximide, an inhibitor of protein synthesis, the ability of red beet root tissue to absorb potassium, sodium and chloride gradually declines and the uptake of each ion is affected at different rates. These observations are consistent with the view that ion transport involves turnover of protein (Sutcliffe, 1962).

3·5 Accumulation in the cytoplasm and vacuole

Ions transported across the plasma membrane enter the cytoplasm of the cell. Within the cytoplasm the ions may be rapidly utilized in the synthesis of organic cell constituents. Anions such as phosphate, sulphate and nitrate are required in large amounts, and this accounts for their macronutrient status. Some cations are also incorporated into cell materials and others play their role as cofactors in enzymically-controlled metabolism as explained in Chapter 2.

Some ions are accumulated within cytoplasmic organelles. Chloroplasts and mitochondria isolated from cells have been found to have an ionic composition which differs considerably from that of the bulk cytoplasm, and in addition they are capable of selectively accumulating ions from a bathing medium. There are a variety of other structures, such as Golgi bodies, as well as smaller aqueous vesicles associated with the endoplasmic reticulum into which ions may be transferred.

Those ions which are not incorporated into cytoplasmic organelles may remain within the cytoplasm as free ions or may become bound to charged molecules at anion and cation exchange sites. These ions may then be transported through the symplasm (see 4.3) or may be transferred across the tonoplast and released within the vacuole.

It is possible that ions are transported across the inner membrane, the tonoplast, into the vacuole by a reversal of the mechanism which is operative at the plasmalemma. Microvesicles which have accumulated ions may fuse with the tonoplast discharging their contents into the vacuole by reverse micropinocytosis. BUVAT (1963) has suggested that dilatation of parts of the endoplasmic reticulum may give rise to small vesicles which later merge with the central vacuole. As a result of active transport across the plasmalemma and the tonoplast ions are accumulated within the vacuoles of plant cells. The analyses of the sap of *Nitella*, a fresh water alga, and *Valonia*, a marine alga, already referred to above (p. 7) give a clear demonstration of both the accumulation and selective properties of the ion uptake mechanism in these plants.

3.6 Factors affecting ion uptake

The active uptake of ions by plant cells is sensitive to a wide variety of external factors (cf. Chapter 1). Of greatest significance are those which affect the energy supply, those which affect growth, and those which concern the composition and concentration of the external medium.

The energy supply may be affected by factors such as oxygen availability, carbon dioxide level, temperature, light and metabolic inhibitors. The metabolic component of ion uptake is inhibited by the absence of oxygen in aerobic organisms. Phosphate uptake by excised barley roots increases with increasing oxygen concentration over the range 0–3% and is unaffected over the range 3–100% when the total gas pressure is maintained at one bar. In the absence of carbon dioxide, photosynthesis is prevented and ion uptake is inhibited as the respiratory substrate becomes depleted. Carbon dioxide, and also bicarbonate ions at high concentrations, inhibit ion uptake as well as other physiological processes by a direct effect on metabolism.

In general, the rate of ion uptake increases with increasing temperature up to about 40° C after which it is progressively reduced due to the denaturation of enzymes. In addition the membranes become more permeable at high temperatures and a greater passive leakage of ions occurs. At low temperatures uptake is decreased by a diminution of active transport through reduced energy supply and by an increase in the resistance of the cell membranes.

A direct effect of light has been demonstrated for chloride uptake by *Nitella* and this is dependent on photo-system II of photosynthesis. Other effects of light are indirect, with visible light supplying the energy for uptake in green plants through the production of photosynthates. Ultraviolet light inhibits ion uptake especially at wave lengths which are absorbed by ribo-nucleic acid (RNA) and may cause leakage due to the destruction of lipo-protein complexes in the membrane.

As active transport is dependent on metabolism it is not surprising that ion uptake is sensitive to all metabolic inhibitors. Particularly effective in inhibiting ion uptake are respiratory inhibitors, such as carbon monoxide, which inhibit electron transport and those such as dinitrophenol (DNP) which uncouple ATP production from the electron transport process (cf. Section 3.4). Inhibitors of protein synthesis, such as chloramphenicol and cycloheximide can reduce ion uptake without any appreciable effect on oxygen consumption although in some instances they may also act as uncouplers and so stimulate oxygen uptake.

Growth is stimulatory to ion uptake in a number of different ways. Accompanying growth is the synthesis of new carrier molecules as well as the incorporation of additional inorganic ions into cell constituents, particularly in those cells which are synthesizing protein. Cell expansion results in a greater surface area of membrane across which ion uptake can occur. When growth slows down, the internal ionic concentration rises and ion uptake declines apparently through direct inhibition of the uptake mechanism.

The surface area across which ions are transported is obviously an important factor in the uptake of ions. In general, small cells which have a high surface to volume ratio absorb ions more rapidly *per unit volume* than do large cells. As might be expected there is also a close relationship between root surface area and ion uptake.

The internal ionic concentration of a cell exerts a marked effect on ion uptake, accumulation declining as internal concentration rises. Ion uptake can be increased in a high salt tissue by reducing its turgidity, an observation for which there is no adequate explanation. Plants grown on a low-salt medium possess a greater capacity for ion uptake than those grown on a normal or high-salt medium. This effect is partly attributable to the high sugar content of low-salt tissues.

In addition to its influence on the availability of ions (Fig. 2–2, p. 20) pH has two major effects—competition and injury. A low pH is believed to reduce cation uptake by competition between hydrogen ions and the substrate cations for sites on a carrier. At high pH values hydroxyl or bicarbonate ions might compete with substrate anions, thus reducing anion uptake. Acidity or alkalinity therefore has a profound influence on the relative absorption of anions and cations. At high pH where the absorption of cations is favoured the discrepancy between cation and anion absorption is balanced by greater accumulation of organic anions within the tissue. The organic acid is synthesized by utilizing carbon dioxide or bicarbonate ions taken up from the medium. At pH values outside the physiological range the ion uptake mechanism is damaged, probably by disruption of membranes.

As discussed in Section 3.3 external concentration has a direct influence on ion uptake and the absorption isotherm is hyperbolic for most ions (Fig. 3–5a). The absorption of one ion may be influenced by the presence of another ion in the medium. The classical example of this is the effect of calcium on the uptake of potassium and chloride. The uptake of potassium and chloride is promoted by calcium up to a certain concentration level and then decreases with further increases in calcium ion concentration. This reflects a requirement for the presence of calcium in plant cell membranes. There are a number of cases of competition between similar ions such as the alkali cations (Na^+ K^+, Cs^+, Li^+ and Rb^+) in which the uptake of one is generally reduced when the concentration of another is increased. This probably reflects competition for uptake sites on a carrier with a low specificity. The presence of metabolically-important anions, such as phosphate and nitrate, often stimulates the uptake of other ions, presumably through an effect on metabolism. Ultimately absorption is dependent on growth and this depends on the availability of all essential elements.

4 Ion Uptake by Plants

4.1 Uptake from the soil

The concentration of free ions in the soil solution is generally low, with the major portion of the cations absorbed on to the negatively-charged sites of clay micelles and organic materials in the soil. Of the anions, nitrate and sulphate usually occur as free ions in the soil solution while phosphate is firmly bound to soil particles and is only found at very low concentrations in the soil solution (Fig. 2–1, p. 15). Anions are absorbed by plants almost entirely from the soil solution, whereas cations may be exchanged directly between roots and soil particles by the process of *contact exchange* (Fig. 4–1a). According to the contact exchange hypothesis cations may be transferred from clay colloids on to roots without appearing as free ions in the soil solution. This is a result of the oscillation

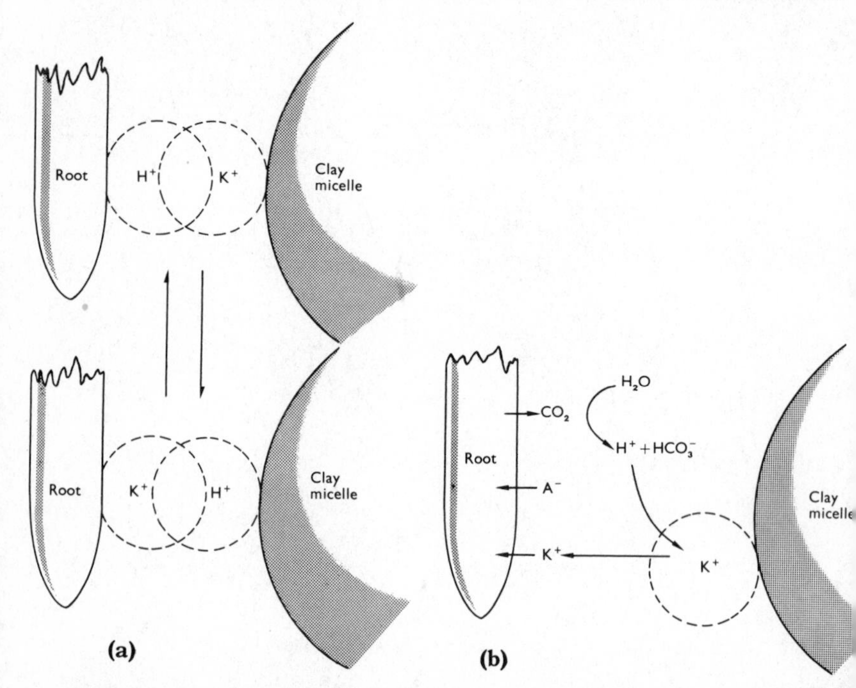

Fig. 4–1 (a) Diagrammatic representation of the contact exchange hypothesis. **(b)** Diagrammatic representation of the carbonic acid exchange hypothesis

of adsorbed ions within a small volume of space. When the oscillation volume of an ion overlaps that of another ion then an exchange of ions may occur.

Not all investigators accept the concept of contact exchange for cation uptake from the soil. Another possibility is that hydrogen ions exchange with the cations absorbed on to soil particles, the hydrogen ions being provided by the carbonic acid formed from respiratory carbon dioxide within the soil. The cation thus released then diffuses through the soil solution to the root surface where ion absorption takes place (Fig. 4–1b).

It is likely that in many soils both contact exchange and carbonic acid exchange occur, the relative contribution of these two processes depending on such variables as soil water content, pH and exchangeable cation content.

Initial entry of ions into the root is mainly through the first few centimetres behind the tip. This includes the region of most active metabolism and extension growth and also the region in which root hairs are most numerous. Beyond this region the older tissue becomes progressively suberized and relatively impermeable, although some absorption of nutrients does occur through these suberized regions of the root.

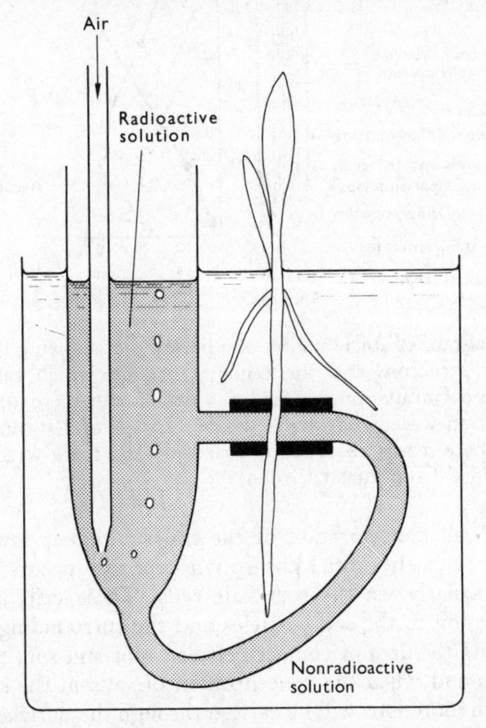

Fig. 4–2 Apparatus to supply radioactive tracers to a specific location on a root. (After WIEBE and KRAMER, 1954)

The region of maximum accumulation does not necessarily coincide with the region of the root through which nutrients enter and pass to the shoot. This latter region may be identified by supplying radioactive isotopes to one segment of a root while the remainder of the root system is immersed in a non-radioactive nutrient solution (Fig. 4–2). Using this technique KRAMER and co-workers have obtained the pattern for both the accumulation and the translocation of a number of ions along the length of the root of different plant species (Fig. 4–3).

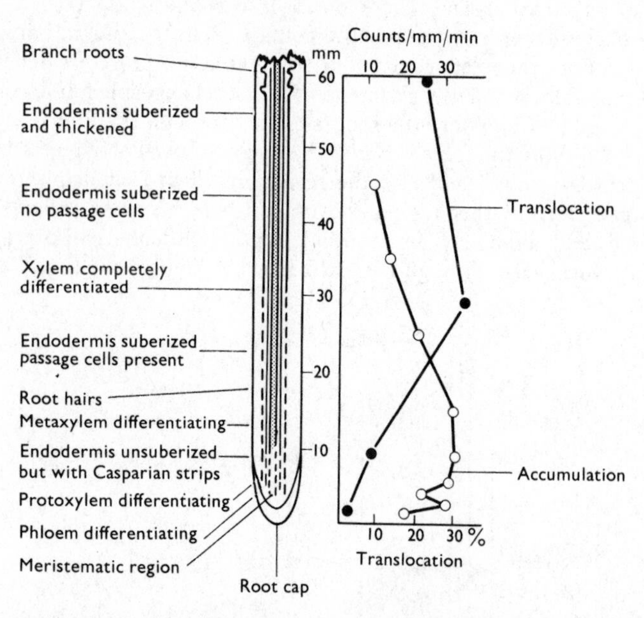

Fig. 4–3 Diagram of apical region of a barley root showing the relationship between root structure and the regions through which salt is absorbed. The curve for accumulation is based on data of WIEBE (1956) and the curve for translocation which represents the percentage of ^{32}P translocated from the region where it was absorbed is based on data from WIEBE and KRAMER (1954) (Modified from KRAMER, 1969)

When the salt concentration of the soil solution is low and at slow transpiration rates the initial absorption probably occurs near the root surface, particularly via the root hair cells. These cells are favourably placed in relation to the soil particles and the surrounding soil solution, thus increasing the area of contact between root and soil. At high transpiration rates and when the concentration of salts in the soil solution is relatively high more ions will be carried through the surface layers of the root and into the cell wall network of free space within the cortex, thus increasing still further the surface area for uptake.

4.2 Transport in the free space

The concept of free space has been outlined in section 3.2. In the case of uptake of salts by the root the free space provides a continuum between the soil solution and all the cells of the root cortex. Figures 4–4a and 4–5a

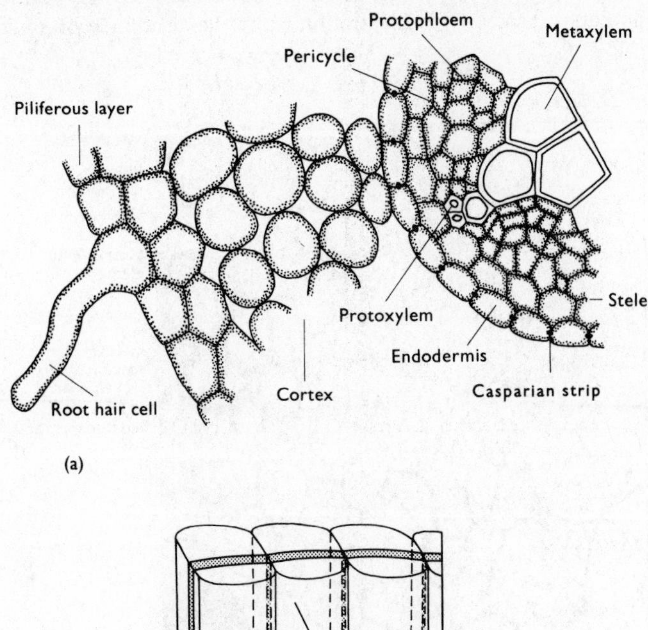

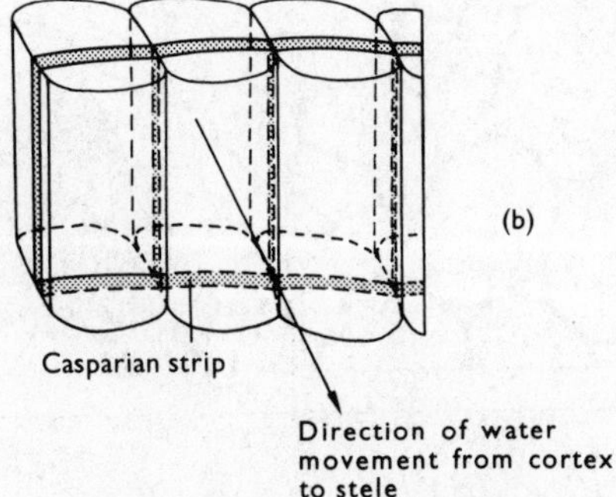

Fig. 4–4 (a) Transverse section of a root. (After PRIESTLEY, 1920). **(b)** Diagram of three endodermal cells showing Casparian strips in transverse and radial walls, but not in tangential walls. (After ESAU, 1960)

show transverse sections of young dicotyledonous roots. As it can be readily demonstrated that roots have a water-free space of at least 10% it is evident that the surface layer of the cortex does not constitute a major barrier to diffusion. The soil solution can penetrate through the cortex as far as the endodermis (cf. Section 4.4) via the cell walls without crossing any membrane or high resistance barrier *en route*. This means that each cortical cell is bathed in a medium very similar to that which surrounds the root and that the outer membranes of these cells, the plasmalemmae, are in direct contact with this solution. This results in a vast surface area over which ion accumulation into the cortical cytoplasm may take place. In this way the cortical cells can remove the bulk of the ions from the solution which is moving across the root in response to the gradient of water potential created by transpiration or by the accumulation of solutes

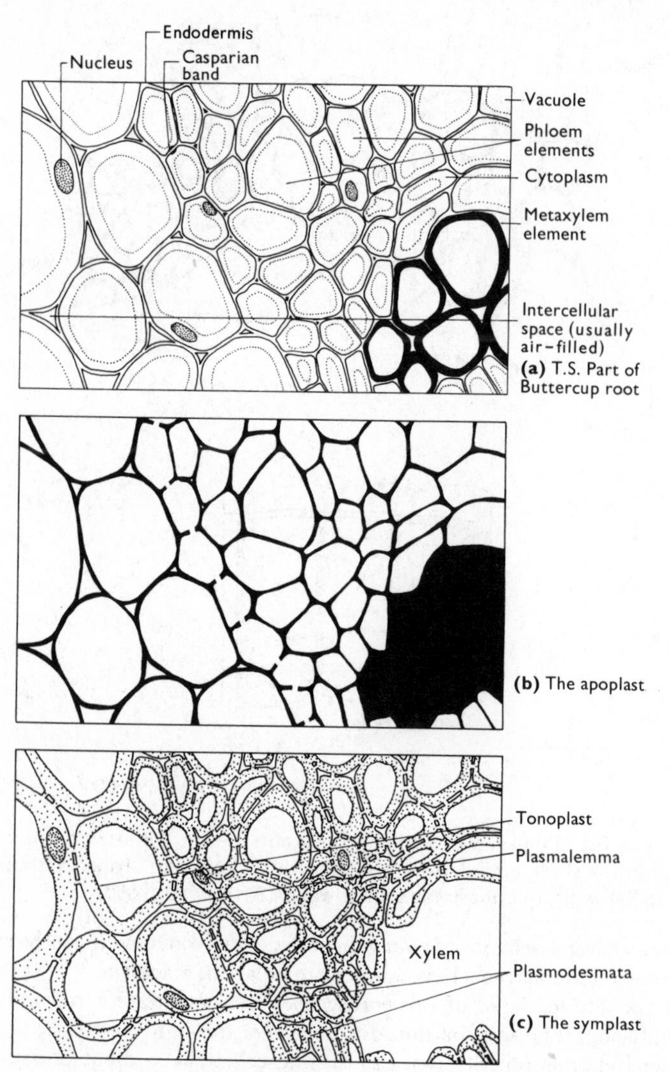

Fig 4–5 TS part of buttercup root **(a)** illustrating apoplast **(b)** and symplast **(c)** components. (After BARON, 1967)

in the xylem sap (Section 4.5). The cortical cell walls have a Donnan phase in which both anions and cations may become bound and some of these together with ions from the water free space are transferred into the cytoplasm.

Those parts of a tissue external to the limiting outer membrane which comprise the free space of the tissue are often referred to as the *apoplast* or *apoplasm* and the above movement of water and ions across the cortex as movement through the apoplastic pathway. As illustrated in Fig. 4–5b there is a break in the apoplastic pathway at the endodermis (Section 4.4).

4.3 Transport in the symplasm

The symplast of the root is illustrated in Fig. 4–5c where it can be seen that it is composed of the cytoplasts of the individual cells, which are joined together by bridges of cytoplasm, the *plasmodesmata*, thus forming a three-dimensional cytoplasmic continuum or network. It is through this continuum that accumulated ions move radially across the cortex and endodermis into the stele, while photosynthetic products from the leaves (sugars, etc.) move in the opposite direction to provide respiratory substrate for the cortical cells. It is generally believed that movement of ions in the symplasm is diffusional through the aqueous phase of the cytoplasm, with cyclosis or cytoplasmic streaming aiding this ion movement. Alternatively, ions may be adsorbed on to proteins or contained in vesicles which are carried along by cytoplasmic streaming. Inhibition of this cyclosis within the cortical cells is known to inhibit the transport of ions across the root.

The driving force for radial diffusional transport is believed to be a declining gradient of concentration from the outer to the inner cells of the root. Little or no longitudinal transport takes place within the cortical symplasm. This has been shown in experiments in which isotopes were fed to only one part of the root, using an apparatus such as that illustrated in Fig. 4–2. Even after several hours, movement in the cortex is limited to a few millimetres on either side of the zone of application of the labelled ion and this could be due to diffusion in the cell walls.

During their passage through the symplasm some ions are transported across the tonoplasts and deposited in the vacuoles of the cortical and stelar cells. Some of them may subsequently be released again and transported to the shoot. Those ions which are destined for long distance transport within the plant are ultimately released within the stele and enter the xylem vessels. The mechanism of this release is as yet unresolved, and investigators are divided about whether there is an active secretion or a passive leakage of ions within the stele. Release of ions from the symplast may take place predominantly from the xylem parenchyma cells surrounding the xylem vessels, which have been shown to accumulate potassium ions to a greater extent than other living cells in the stele. Alternatively they may be released into the apoplasm anywhere within the stele and and be carried from there passively in the transpiration stream into the xylem vessels

4.4 The endodermis

It has been mentioned in previous sections that the endodermis provides a boundary to the free space of the root. Back diffusion into the cortex of ions released in the apoplasm within the stele is prevented by the endodermis which in addition provides an effective barrier to the passive movement of water through the apoplast between these two tissues. Such movement is prevented by a suberin-impregnated region in the radial and transverse walls of the endodermis, the Casparian strip (Fig. 4–4b). Suberin renders the cell wall impervious to water and to solutes and thus movement across the endodermis must be through the cytoplasm. As the root matures the endodermal cells become thickened further which gives the endodermis a characteristic toothed appearance in transverse section. Often the endodermal cells nearest to the protoxylem do not have this secondary thickening and are referred to as *passage cells*. However, the walls of the passage cells are impregnated by a Casparian strip and so are not gaps in the endodermal barrier, as suggested in some elementary texts. The secondary thickening is traversed radially by large plasmodesmata and therefore probably does not reduce the radial cytoplasmic path.

Practical demonstrations that the endodermis is a barrier have been obtained by two different methods—autoradiography and measurements of free space. The technique of tissue autoradiography allows radioactive isotopes to be located within sections of tissue. The labelled tissue is placed on photographic film and the emission from the isotope causes a darkening of the film. When roots are exposed to labelled sulphate for a short time under conditions which prevent active movement (e.g. at low temperature) and the sulphate then precipitated *in situ* in the form of an insoluble salt, such as barium sulphate, autoradiography reveals that movement is only as far as the endodermis. Free space measurements have shown that the free space value for excised roots is higher than that for intact roots. This demonstrates that a radial barrier has been broken in the excised roots thus making an additional portion of the root directly accessible to the external solution. The free space value for the cortex alone is the same as that for the intact root, implying that this radial barrier is located at the endodermis.

4.5 Root exudation

A commonly observed phenomenon is the exudation of a watery liquid which occurs from the cut stumps of plants. If a healthy, well-watered plant, such as tomato, is de-topped and a glass tube connected to the stump with a rubber sleeve, it is possible to collect the solution which exudes from the cut xylem vessels. This exudation is the result of a positive hydrostatic pressure which develops in the xylem. Most investigators now accept that root pressure is a purely osmotic phenomenon, although claims have been made at various times that exudation is a result of active water transport. Analyses of the exudate reveal that the concentration of an ion in the exudate is often many times greater than that in the external solution. The reduced solute potential (increased osmotic pressure) within the xylem causes a gradient of decreasing water potential across the root from outside to inside. The symplasm and endodermis together provide a semipermeable membrane across which water movement occurs causing a positive pressure to develop in the stele resulting in a flow of solution up the xylem conduits.

The solute potential is lowered by the release of ions from the symplasm within the stele. As indicated in Section 4.3 there is currently some controversy over the question of how these ions are released. Some investigators believe that the plasmalemmae of the stelar cells are leaky and that ions diffuse out passively from the stelar symplast. The leaky nature of the stelar tissues is attributed to a supposedly diminished metabolism within the centre of the root caused by a low oxygen and high carbon dioxide concentration. The alternative view is that ions are secreted by the stelar parenchyma cells, a process which would involve some form of active pump operating at the plasmalemma of the stelar symplast.

When transpiration is occurring ions are still moved into the stele but their concentration may fall because the water is moving in faster. The pressure within the xylem often becomes negative under transpiring conditions, that is, a tension develops. The effects of water movement on the uptake and transport of ions within the plant are considered in the following section.

In the intact plant the external manifestation of root pressure is the extrusion of droplets of fluid from the leaves of some plants. This phenomenon, termed *guttation*, only occurs when transpiration is minimal. The volume of fluid exuded by a guttating plant varies from a few drops on a grass blade overnight to the prodigious 200 ml recorded from a single leaf in one day for the Indian taro plant, *Colocasia antiquorum*. The volume and composition of guttation fluids are extremely variable, ranging from almost pure water to a solution of ions with a solute potential of -1 bar or lower. The physiological importance of guttation is in doubt, but it ensures that solutes are carried from the roots to the leaves of plants through the xylem in the absence of transpiration.

4.6 Relationship between water and ion transport

It is well established that under certain conditions an increased water uptake is accompanied by increased absorption and transport of ions to the shoot. It remains uncertain whether transpiration is altering the rate of an active transport process directly or the effect is indirect due to an increased supply of ions at the root surface with increased transpiration. Alternatively, the removal of ions from within the stele by transpiration will lower the concentration there and cause the active uptake mechanism to speed up.

It is probable that in older roots, where the endodermis is pierced by branch roots, leaks may exist in the barriers to ion movement and a mass flow of the external solution may take place across the root into the xylem. This component of ion uptake will be directly proportional to the transpiration rate.

4·7 Foliar absorption of ions

Many aquatic plants obtain the bulk of their nutrient supply through the surface of their leaves, and the same is probably true of aerial epiphytes. However, all plants will readily absorb mineral nutrients applied to their aerial surfaces and this fact is exploited in modern agriculture. Foliar fertilization has been successfully used for the application of urea nitrogen, phosphorus, magnesium and various micronutrients to pineapple, sugar-cane, citrus and various forest trees. In general foliar application is an adjunct to, rather than a substitute for, soil fertilization. Not only mineral nutrients but herbicides, antibiotics, growth regulators, systemic insecticides and fungicides are applied in foliar sprays.

The pathways of a penetrating ion may be through the stomata or leaf cuticle or both. The use of an ionic surfactant or detergent to lower the surface tension of a foliar-applied solution markedly increases absorption through the stomata. Cuticular penetration has been studied and it appears that the cuticle is often the main route of entry. Plasmodesmata-like strands occur in the outer cell walls of leaves, often extending up to and just beneath the cuticle. These strands, termed *ectodesmata*, may provide pathways for the transport of substance into and out of leaves.

In general, nutrients penetrate more rapidly into young than into mature leaves, which may reflect structural differences in the epidermis and cuticle. Differences in the degree of metabolic activity of the leaves may also be involved. Some of the ions absorbed after foliar application are retained, especially in young leaves, the remainder being exported through the phloem (see Section 5·3).

5 Distribution of Ions

5.1 Patterns of ion distribution and circulation

Once ions and other solutes have reached the xylem ducts they are transported upwards in the transpiration stream to the shoot where they are distributed throughout the plant. Ions which have been deposited in leaves during the growing season are moved out as the leaves senesce prior to abscission and are then translocated to storage tissues in the stem or root (see Section 5.4). However, within vigorously-growing plants prior to any senescence there is a general redistribution of mobile ions

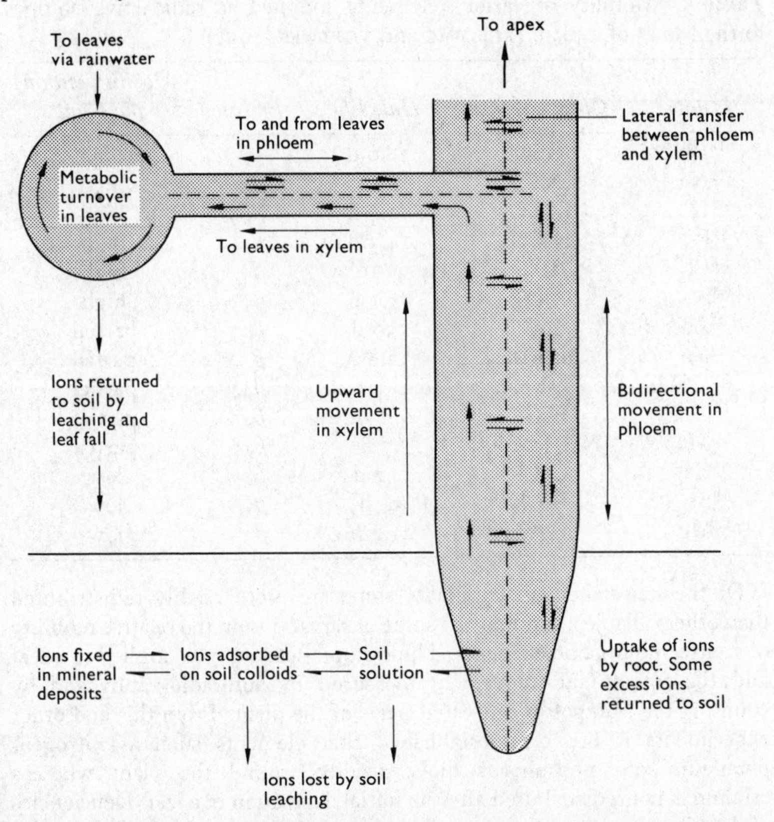

Fig. 5–1 Diagram of the circulation of minerals within a plant and their return to the soil by leaching and leaf fall. (Modified from KRAMER, 1969)

out of older leaves to young leaves and reproductive areas. This circulation of elements is shown diagrammatically in Fig. 5–1. Generally movement takes place in the vascular tissues, the initial upward movement being in the xylem while the redistribution takes place in the phloem. There is a considerable amount of lateral transfer of solutes between phloem and xylem which sometimes results in the simultaneous upward movement of an element in the xylem and its downward movement in the phloem.

This redistribution is in response to, and probably controlled by, the relative metabolism of different areas of the plant, ions tending to move towards metabolically-active regions ('sinks') such as apical meristems, young leaves, and reproductive structures perhaps in response to the increased hormone production by these regions (see p. 55). The way in which this internal redistribution is controlled by these hormones is currently the subject of investigation and as yet the mechanism is not known.

Table 5 Mobility of various elements supplied as radioactive isotopes to the leaves of a plant (BUKOVAC and WITTWER, 1957)

Isotope	*Chemical form*	*Half life*	*Emission*	*Mobility within the plant*
^{86}Rb	$RbCl$	18.6 d	β, γ	high
^{22}Na	$NaCl$	2.6 y	β, γ	high
^{42}K	K_2CO_3	12.4 h	β, γ	high
^{32}P	H_3PO_4	14.3 d	β	high
^{36}Cl	HCl	4×10^5 y	β	high
^{35}S	SO_4	87.1 d	β	high
^{65}Zn	$ZnCl_2$	250 d	β, γ	partial
^{64}Cu	$Cu(NO_3)_2$	12.8 h	β, γ	partial
$^{52-54}Mn$	$MnCl_2$	6.2–310 d	β, γ	partial
$^{55-59}Fe$	$FeCl_3$	46 d–2.94 y	β, γ	partial
^{99}Mo	$(NH_4)_2MoO_4$	67 h	β, γ	partial
^{45}Ca	$CaCl_2$	152 d	β	low
^{89}Sr	$SrCl_2$	54 d	β	low
^{28}Mg	$MgCl_2$	21.4 h	β	low

Of the ions taken up by plants some are more readily redistributed than others. Table 5 summarizes some observations on the relative mobility of fourteen inorganic elements. The isotopic solution was applied to a leaf and the uptake and movement measured by autoradiography and by counting the isotopes in untreated parts of the plant. From this and other experiments it has been established that elements such as nitrogen, potassium and phosphorus move readily around the plant whereas calcium is not redistributed after its initial deposition in a leaf. Deficiencies of mobile elements appear in the older leaves first, while deficiencies of non-mobile elements appear in the youngest leaves first (see Chapter 2).

Phosphorus is highly mobile and is probably continuously circulated within the plant. A given phosphorus atom may make several complete circuits of a plant within a day (BIDDULPH, 1959). The vital role of phosphorus in respiration, photosynthesis, starch, nucleic acid, fat and protein synthesis, means that phosphorus is being incorporated and released continuously at the various points within the plant where any one of these processes takes place.

The rapid uptake of sulphur into metabolic compounds prevents its ready circulation within the plant, although this element is freely mobile within the translocation system. Sulphur, an important component of proteins, is ultimately fixed in those regions of the plant in which protein synthesis is dominant, such as the stem and root apex and the young leaves. Proteolysis eventually liberates this fixed sulphur which is then translocated to other regions of protein synthesis. Thus a very slow turnover and circulation of sulphur occurs within the plant, which led some early investigators to the view that this element was immobile.

Calcium is one of the least circulated elements within the plant. It is readily taken up and carried in the transpiration stream to various parts of the plant. Some cases of calcium transport in phloem e.g. into developing fruits (5.4) and out of cotyledons have been reported (see 5.5). However, the relative immobility of calcium in the phloem usually prevents circulation of this element after its initial deposition from the transpiration stream.

Prior to abscission of leaves there is a withdrawal of some mineral nutrients, among which are nitrogen, potassium, phosphorus, sulphur and chlorine. Under certain conditions, magnesium and iron are also withdrawn to some extent prior to abscission. Those elements which remain within the leaf, and are lost to the plant on leaf fall, include sodium, calcium, boron, manganese and silicon.

Exact evaluation of the extent to which minerals are recirculated in plants is complicated by the fact that considerable amounts of some elements are leached from leaves as a result of rain or dew. In addition some of the excess ions not used in metabolic processes may return to the roots and leak out into the surrounding medium.

5.2 Transport in the xylem

The upward movement of ions in the xylem tissues has been demonstrated in a number of different ways. Ringing experiments have shown that the upward movement of ions is unimpeded by the removal of the phloem tissue. Upward movement in some instances follows the same spiral patterns as reported for the movement of dyes in the xylem. Radioactive tracers supplied to the roots have been reported to move at rates of up to 60 metres per hour in rapidly-transpiring trees, while in herbaceous plants tracers often reach the shoot tips in less than one hour (KRAMER, 1969).

The relationship between transpiration and ion uptake (cf. Section 4.6) is also consistent with the concept that ions are carried upward in the transpiration stream (see Plate 4).

Relatively large amounts of dissolved ions have been detected by direct analysis of extracted xylem sap. The composition and concentration of the xylem sap varies with the species, the season and the time of day. The ionic composition and concentration of the rooting medium also exerts a profound effect on the xylem sap.

Not all mineral transport in the xylem occurs as inorganic ions. Most of the nitrogen is transported in the form of amides, amino acids and ureides. Small amounts of organic sulphur and phosphorus compounds have also been reported. The occurrence of these organic compounds in xylem sap implies that the ions must be involved in metabolic reactions as they cross the root. The mechanism by which these organic compounds enter the xylem is unknown.

The movement of certain ions through the xylem is restricted by their tendency to precipitate. Iron may be precipitated as ferric phosphate in the root cortex or the xylem to such an extent that the leaves can become deficient. Similarly, zinc may be absorbed in considerable quantities on the walls of xylem vessels in the grape vine. Thus deficiencies of certain elements may occur even though adequate quantities are being absorbed by the roots.

In addition to the upward translocation of ions in the xylem, there is also a considerable lateral movement between the vascular tissues. This movement normally takes place between xylem and phloem across the cambium and it has been postulated that the cambial tissue may regulate the amount of ions carried upwards in the transpiration stream. Active accumulation of ions by the cambial cells may discriminate between the various ions in the xylem and transfer them laterally to the phloem and similarly from phloem to xylem. For instance, if an ion is present at a low concentration in the phloem it may be actively accumulated in the cambium from the xylem sap and its lateral translocation into the phloem enhanced. If, however, an equilibrium existed for an ion between the phloem and the cambium then the upward passage of that ion in the transpiration stream would be unimpeded.

5·3 Transport in the phloem

As indicated above there is a considerable amount of ion movement through the phloem tissues of plants. Analyses of extracted phloem sap have shown that considerable quantities of some elements are present in the phloem in both inorganic and organic forms (Table 6). These minerals are those which are exported from leaves prior to abscission, and, as discussed above, those which are circulated and re-utilized within the plant. The almost complete absence of inorganic nitrogen is a feature of phloem exudates. However, as can be seen in Table 6, a considerable quantity of amino acids is transported in the phloem.

Table 6 The composition of the exudate obtained from incisions made in the bark of *Ricinus* plants (from HALL and BAKER, 1972)

	concentration	
	mg per ml	*m eq per l*
Dry matter	100–125	
Sucrose	80–106	
Reducing sugars	Absent	
Protein	1.45–2.20	
Amino acids	5.2 (as glutamic acid)	35.2 (mM)
Keto acids	2.0–3.2 (as malic acid)	30–47
Phosphate	0.35–0.55	7.4–11.4
Sulphate	0.024–0.048	0.5–1.0
Chloride	0.355–0.675	10–19
Nitrate	Absent	
Bicarbonate	0.010	1.7
Potassium	2.3–4.4	60–112
Sodium	0.046–0.276	2–12
Calcium	0.020–0.092	1.0–4.6
Magnesium	0.109–0.122	9–10
Ammonium	0.029	1.6

The movement of ions in the phloem has been demonstrated in studies using radioactive tracers which have shown that the outward movement of ions from a leaf is both upwards and downwards in the phloem with lateral transfer of ions between phloem and xylem. The question of whether this bidirectional movement within the phloem may take place within an individual sieve tube or in separate channels is a controversial one which has not yet been resolved.

The mechanism by which these minerals are moved through the phloem is presumably the same as that by which sugars are translocated. This complex subject has been dealt with elsewhere in this series (RICHARDSON, 1969, *Translocation in Plants*), and we will restrict our observations to those which are pertinent to ion transport in the phloem tissue. It has been reported that different substances are translocated at different rates through the phloem. For instance BIDDULPH and CORY (1957) demonstrated that ^{14}C labelled sucrose moved considerably faster (107 cm h^{-1}) than either tritiated water or ^{32}P labelled phosphate (86.4 cm h^{-1}) when applied simultaneously to bean plants. This result suggests that these substances may be transported independently, but the very rapid exchange of tritium (^3H) with stable hydrogen atoms and the rapid fixation of phosphate in metabolism makes the interpretation of such a result difficult. The time taken for the different labelled substances to enter the phloem may also vary, and thus affect the apparent rate of transport.

It has been suggested that potassium is the ion which drives translocation via an electro-osmotic mechanism (SPANNER, 1958). The analysis presented in Table 6 indicates the presence of a considerable quantity of potassium in the phloem, a level which is probably too high for optimal electro-osmotic efficiency. Furthermore, a variety of anions are present within the phloem sap while the proposed cation-driven electro-osmosis requires conditions inimical to anion transport.

A high ratio of magnesium to calcium is characteristic of phloem saps and it has been speculated that this feature may be important if the translocation mechanism involved contractile protein. Streaming in algal cells such as *Nitella* is promoted by magnesium and depressed by calcium. The motive force for this streaming is generated by bundles of microfilaments each 5 nm in diameter. Much of the protein found in the sieve tube is in the form of microfilaments, termed *p-protein*, and it has been suggested that these filaments may generate a force of sufficient magnitude to drive the phloem sap along the sieve tubes. However, contractility in such fibres remains to be demonstrated, and p-protein does not occur in all plants.

5·4 Accumulation in storage organs

During the development of storage organs, such as bulbs, rhizomes, corms, tubers, fruits and seeds, materials—including inorganic ions, are diverted from various parts of the plant towards the regions of active growth, and the levels of such elements as nitrogen, phosphorus and potassium in the leaves and stems may actually fall. Because there is very little transpiration from storage organs, sap flow in the xylem is slow and so the bulk of the solute movement must occur in the phloem. MASON and MASKELL (1931) showed that the amounts of certain elements including nitrogen, phosphorus and sodium, as well as of carbohydrate, translocated into cotton (*Gossypium hirsutum*) bolls is vastly in excess of the quantity carried in the xylem. On the other hand, they found that accumulation of calcium was proportional to the calcium content of tracheal sap. Experiments in which a section of the stem supplying a storage organ is killed by steam ('steam-girdling') have demonstrated that the movement of potassium, nitrogen, phosphorus and other phloem-mobile elements (Table 5), but not of calcium, depends on living tissues, and it has been confirmed by radio-autography that such movement occurs in the sieve tubes. Using the electron micro-probe analyser, calcium has been detected in the sieve tubes of the fruit stalk in the garden pea (*Pisum sativum*) and it is concluded that some transport of calcium into the pod occurs through the phloem (see 5.5 below). In contrast, there is very little transport of labelled calcium into the fruits of the pea-nut (*Arachnis hypogaea*) when the roots are immersed in a solution containing radioactive calcium. Pea-nut fruits normally develop underground on special structures, the *gynophores*, produced after fertilization of the flowers and they grow down into the soil from which they are able to absorb calcium direct.

Accumulation of substances in a fruit occurs only in those parts which are growing at any particular time. Labelled phosphorus supplied to the roots of tomato plants has been detected in the pulp of those fruits which are still enlarging, whereas in those which have reached full size, phosphorus only accumulates in the seeds. The physiological basis of the 'sink' effect which causes materials to move through the phloem towards regions of active growth is still not adequately explained. A lowering of the level of soluble substances in the vicinity of the vein endings as a consequence of their utilization in synthetic processes is likely to be an important factor. Growth hormones, such as I.A.A. cytokinins and gibberellins, probably exert their effects mainly by controlling these processes, but a more direct influence on translocation cannot yet be excluded. When materials are transferred from one generation to another as during development of seeds and in the transport of nutrients from host to parasite through haustoria, special secretory cells, the so-called *transfer* cells, may be involved but nothing is yet known about their mode of action or the mechanism of their regulation.

Release of substances by the supplying organ (the 'source') involves an increase in the permeability of cell membranes, which in the case of senescing leaves is associated with development of activity of hydrolytic enzymes, such as proteases and ribo-nucleases. These processes are under hormonal control and can sometimes be reversed, e.g. by the application of cytokinins. Fruiting, especially in annual plants in which a large proportion of the total dry matter is transferred into the seed, sometimes increases the permeability of root cells to such an extent that ions, such as potassium, are released in considerable quantities from vacuoles into the external medium.

5·5 Ion distribution in germinating seeds

Materials accumulated in the storage organs of a developing seed are re-mobilized and transported into the growing embryo during subsequent germination. The rate of depletion of cotyledons or endosperm is controlled in some way by growth of the axis. Conditions which promote growth, such as an optimal temperature, adequate aeration and a supply of inorganic ions in the medium bathing the roots, stimulate the mobilization and transport of materials. Removal of the shoot at an early stage, leaving the cotyledons attached to the roots, does not immediately affect root growth in peas or influence the rate at which potassium is transported into the roots from the cotyledons (SUTCLIFFE, 1962). This indicates that at least in the early stages of growth there is no serious competition between root and shoot for the available nutrients. The distribution of cotyledonery reserves between shoot and root in peas depends on the age of the seedlings and the growth conditions. The percentage of the exported reserves accumulated in the shoot increases with time and this is correlated with cessation of root growth after about two weeks from the beginning of germination, while growth of the shoot continues for up to four weeks, even in the dark. When pea seedlings are grown in total dark-

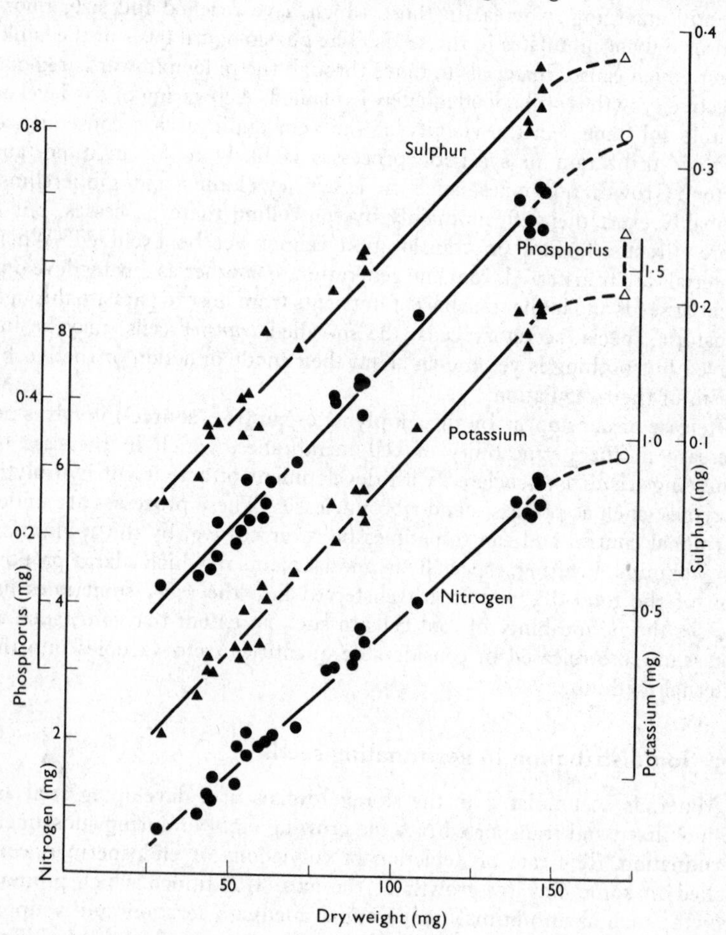

Fig. 5–2 The relationship between dry matter content and the amounts of individual elements in the cotyledons of *Pisum sativum* during a germination period of 4 weeks. The seedlings were grown under a variety of conditions and cotyledons analysed after 1, 2, 3 and 4 weeks by which time they were almost depleted of nitrogen. (GUARDIOLA and SUTCLIFFE, 1972)

ness the cotyledons are depleted more rapidly than when the plants are in light. Dark-grown seedlings also have a higher fresh and dry weight in the early stages of germination, although with the onset of photosynthesis, the dry weight (but not the fresh weight) of light-grown plants eventually exceeds that of plants grown in the dark.

The rates at which different elements are transported out of cotyledons or endosperm varies greatly between elements and in different plants. In general, the movement of nitrogen, phosphorus sulphur and potassium follows closely that of dry matter. It has been found in peas that except during the first week of germination when these elements were exported from the cotyledons relatively more slowly than total dry matter, there was a linear relationship between the transport of individual elements and loss of dry matter (Fig. 5–2). The ratios in which the various elements were exported were the same whether the plants were grown in light or darkness and were independent of the composition of the solution supplied to the roots, even when these treatments influenced the rates of transport. The most likely explanation of this is that soluble materials are released from the storage cells in constant proportions irrespective of the rate of transport. Only a relatively few cotyledon cells may be releasing materials at any one time and this number is controlled in some way by the axis.

Calcium is transported less readily than other elements from cotyledons and in some plants it does not move at all. About 25% of all the calcium in pea cotyledons moves readily into the axis and the rest is immobile. The former fraction is possibly released from phytin during germination and the latter may be associated with cell wall materials. Movement of calcium into the shoots of pea seedlings is reduced by a steam girdle which suggests that transport from the cotyledons occurs in the phloem. About half of the calcium leaving the cotyledons goes to the roots and this movement is unlikely to occur through the xylem because the transpiration stream must flow in the opposite direction.

In contrast to the observations made on peas, the movement of potassium into the axis of etiolated oat seedlings exceeds that of total dry matter in the early stages of germination (Fig. 5–3). The rate falls off rapidly with time as the endosperm becomes depleted. It appears that nearly all the potassium in oat endosperm is readily available and that there is a strong sink in the growing axis which promotes movement as long as supplies are available. A large supply of potassium to the roots reduces the rate of movement of this element relative to others from the endosperm, suggesting that the sink strength of the axis diminishes with increasing potassium content.

The relatively slow movement of nitrogen, phosphorus and magnesium compared with potassium from the endosperm of oat seedlings during early germination is attributable to the fact that these elements are present initially as insoluble materials which must be hydrolysed to soluble substances before transport can occur. The breakdown of starch and

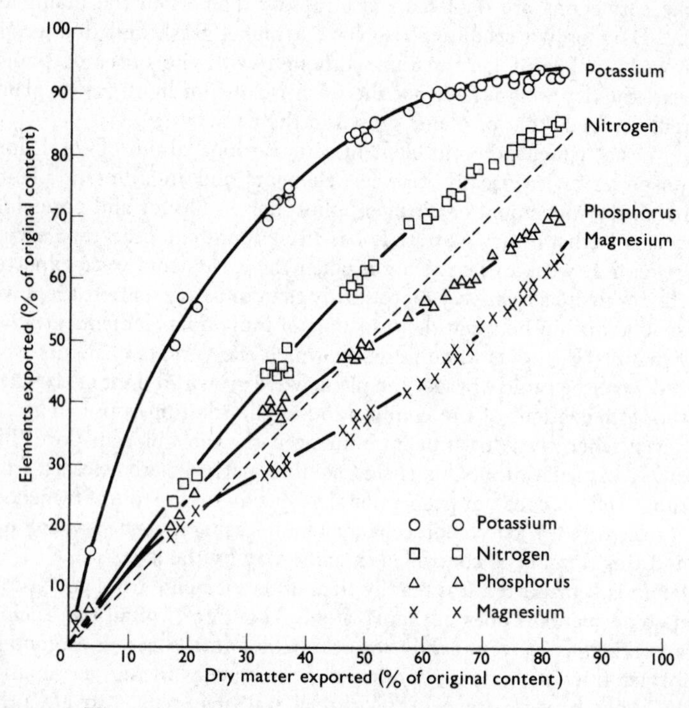

Fig. 5–3 The relationship between the export of dry matter and that of potassium, nitrogen, phosphorus and magnesium from the endosperm of etiolated oat seedlings. The dotted line shows the relationship if dry matter and an individual element are depleted at the same rate. (BASET, 1972)

protein in oat endosperm is closely correlated with the rate of synthesis of hydrolytic enzymes, amylases and proteases which decompose these substances. The growing axis appears to control the movement of materials from the endosperm partly by acting as a sink and partly by regulating the synthesis of enzymes involved in mobilization and thus the availability of elements at the source. It is known that gibberellic acid (GA_3) produced by the embryo instigates enzyme synthesis in cereal grains. Thus we perceive that the germinating cereal grain is an integrated self-regulatory system in which growth of the axis controls, not only the rate of utilization of materials, but also their mobilization.

Further Reading and References

Further reading

EPSTEIN, E. (1972). *Mineral Nutrition of Plants, Principles and Perspectives.* Wiley, New York. A personalized account of mineral nutrition of plants by a leading researcher in the field of salt uptake.

GAUCH, H. G. (1972). *Inorganic Plant Nutrition*, Dowden, Hutchinson and Ross Inc. Stroudsburg, Pa. A sourcebook on plant nutrition with over 2600 references.

HEWITT, E. J. (1966). *Sand and Water Culture Methods used in the Study of Plant Nutrition.* 2nd edn. Tech. Comm. 22. East Malling, Kent. Contains useful information about apparatus, media and methods used in the study of plant nutrition.

HOAGLAND, D. R. (1944). *Lectures on the Inorganic Nutrition of Plants.* Chronica Botanica, Waltham, Mass. Written by one of the pioneers of modern research on salt absorption and still well worth reading.

HOPE, A. B. (1971). *Ion Transport and Membranes.* Butterworths, London. Biophysical approach requiring fairly sophisticated mathematical knowledge despite the author's claim that the treatment is at a rather simple level!

KRAMER, P. J. (1969). *Plant and Soil Water Relationships.* McGraw-Hill. Especially relevant is the chapter on the absorption of solutes, which integrates this process with water absorption.

NOBEL, P. S. (1970). *Plant Cell Physiology.* Freeman, San Francisco. Biophysical and physiochemical approach to ion uptake. Suitable for advanced students.

RUSSELL, E. J. (1974). *Soil Conditions and Plant Growth.* Longmans, London. 10th edn. (revised by E. W. Russell). First published in 1912—a classic on agricultural aspects of plants in relation to soil.

STEWARD, F. C. (ed.) (1959). *Plant Physiology—A Treatise. Volume II Plants in Relation to Water and Solutes.* Academic Press, New York. Chapters on: cell membranes (R. Collander); plants in relation to inorganic salts (F.C. Steward and J. F. Sutcliffe); translocation of inorganic solutes (O. Bidduph).

STEWARD, F. C. (ed.) (1963). *Plant Physiology—A Treatise.* Volume III, *Inorganic Nutrition of Plants.* Academic Press, New York. Mineral nutrition of plants in soils and in culture media (C. Bould and E. J. Hewitt); requirements of essential elements and their interactions in plants (E. J. Hewitt); role of essential mineral elements (A. Nason and W. D. McElroy); microbial activities in soil as they affect plant nutrition (J. H. Quastel).

SUTCLIFFE, J. F. (1962). *Mineral Salts Absorption in Plants.* Pergamon, Oxford. Comprehensive introductory text on salt uptake and transport; now unfortunately out of print, but a new edition is on the way.

TRUOG, E. (ed.) (1951). *Mineral Nutrition of Plants.* University of Wisconsin Press, Madison. Readers comparing knowledge in 1951 with the present day will be surprised how little progress seems to have been made in solving some of the fundamental problems of mineral nutrition.

WALLACE, T. (1961). *The Diagnosis of Mineral Deficiencies in Plants.* H.M.S.O.,

London. Has 312 colour plates showing deficiency and toxicity symptoms in many crop plants.

Reviews of various aspects of Mineral Nutrition and Salt Transport occur regularly in the *Annual Reviews of Plant Physiology*. Annual Reviews Inc., Palo Alto, California, and in a number of symposia.

References

ARNON, D. R. and HOAGLAND, D. R. (1940). *Soil Sci.*, **50**, 463–84.
ARNON, D. R. (1950). *Lotsya*, **3**, 31–8.
BARON, W. M. M. (1967). *Water and Plant Life*. Heinemann, London.
BASET, Q. A. (1972). D.Phil. thesis. University of Sussex.
BIDDULPH, O. (1959). In *Plant Physiology—A Treatise*, Vol. II, ed. F. C. Steward (see Further Reading).
BIDDULPH, O. and CORY, R. (1957). *Pl. Physiol.*, **32**, 608–19
BOWEN, H. J. M. (1966). *Trace Elements in Biochemistry*. Academic Press, London and New York.
BROWNELL, P. F. and CROSSLAND, C. J. (1972). *Pl. Physiol.*, **49**, 794–7.
BUKOVAC, M. J. and WITTWER, S. H. (1957). *Pl. Physiol.*, **32**, 428–35.
BUVAT, R. (1963). *Int. Rev. Cytol.*, **14**, 41–155.
COLLANDER, R. (1941). *Pl. Physiol.*, **16**. 691–720.
CRESSWELL, C. F. and NELSON, HELEN (1972). *Ann. Bot.*, **36**, 771–80.
DAINTY, J. (1969). In *Physiology of Plant Growth and Development*, ed. M. B. Wilkins. McGraw-Hill, London and New York.
ESAU, K. (1960). *Anatomy of Seed Plants*. Wiley, New York and London.
GAUCH, H. G. and DUGGAR, W. M. (1953). *Pl. Physiol.*, **28**, 457–66.
GOLDACRE, R. J. and LORCH, I. J. (1950). *Nature*, **166**, 497–9.
GUARDIOLA, J. L. and SUTCLIFFE, J. F. (1972). *J. exp. Bot.*, **23**. 322–37.
HALL, S. M. and BAKER, D. A. (1972). *Planta (Berl.)*, **106**, 131–40.
HEWITT, E. J. (1963). In *Plant Physiology—A Treatise*, Vol. III, ed. F. C, Steward (see Further Reading).
HOMÈS, M. V. (1963). *Soil Sci.*, **96**, 380–6.
LATIES, G. G. (1969). *A. Rev. Pl. Physiol.*, **20**, 89–116.
LUNDEGÅRDH, H. (1955). *A. Rev. Pl. Physiol.*, **6**, 1–24.
MASON, T.G. and MASKELL, E. J. (1931). *Ann. Bot.*, **45**. 125–73.
MITCHELL, R. L. (1954). *Analyse des Plantes et Problèmes des Engrais Minéraux*. Paris.
PRIESTLEY, J. H. (1920). *New Phytol.*, **19**, 189–200.
REISENAUER, H. M. (1966). In *Environmental Biology*, ed. P. L. Altman and D. S. Dittmer. *Fed. Ann. Soc. Exp. Biol.*, Bethesda, Maryland.
RICHARDSON, M. (1969). *Translocation in Plants*. Edward Arnold, London.
ROBERTSON, R. N. and TURNER, J.S. (1945). *Aust. J. exp. Biol.*, **23**, 63–73.
ROBERTSON, R. N. (1968). *Protons, Electrons, Phosphorylation and Active Transport*. Cambridge University Press.
SPANNER, D. C., (1958). *J. exp. Bot.*, **9**, 332–42.
SPANSWICK, R. M. and WILLIAMS, E. J. (1964). *J. exp. Bot.*, **15**, 193–200.
SUTCLIFFE, J. F. and COUNTER, E. R. (1962). *Nature*, **183**, 1513–14.
TRIBE, M. and WHITTAKER, P. (1972). *Chloroplasts and Mitochondria*. Edward Arnold, London.
TWYMAN. E. S. (1951). *New Phytol.*, **50**, 210–26.
WEISS, M. G. (1943). *Genetics*, **28**, 253–68.
WIEBE, H. H. and KRAMER, P. J. (1954). *Pl. Physiol.*, **29**, 342–8.